Scheduling Residential Construction

For Builders and Remodelers

Thomas A. Love

Home Builder Press®
National Association of Home Builders
1201 15th Street, NW
Washington, D.C. 20005-2800

Scheduling Residential Construction for Builders and Remodelers
ISBN 0-86718-401-9

Printed in the United States of America on recycled paper

Library of Congress Cataloging-in-Publication Data
Love, Thomas A., 1956-
Scheduling residential construction for builders and remodelers / Thomas A. Love.
p. cm.
ISBN 0-86718-401-9
1. House construction—Management. 2. Scheduling (Management)
I. Title.
TH4812.L68 1995
690'.837'068—dc20 94-41539
CIP

For further information please contact—
Home Builder Press®
National Association of Home Builders
1201 15th Street, NW
Washington, DC 20005-2800
(800) 223-2665

4/95 HBP/McNaughton/3,500

Contents

Figures

About the Author

Thomas Love has been teaching scheduling and computers for the Department of Building Science at Auburn University for the past eight years. He has also conducted courses in materials, scheduling, and estimating for the Home Builders Association of Alabama as part of the Graduate Builders Institute. Tom has seven years of construction experience, a bachelor of science from Auburn University, and a master's degree from Colorado State University. He worked as a project manager in both commercial and residential construction, the latter including both single-family and multifamily construction.

Acknowledgments

For their thoughtful reviews of the outline and manuscript, special thanks go to Hunt Dalton, H.D. Homes, High Point, NC; John Geoffroy of Construction Data Control, Inc., Atlanta, GA; Jerry Householder, Chair, Department of Construction, Louisiana State University, Baton Rouge, LA; Professor Bill Huston, Assistant Professor, Department of Building Science, Auburn University, Auburn, AL; Jay S. Newitt, Department of Technology Education and Construction Management, Brigham Young University, Provo, UT; Jonathan Wallick, Wallick Construction Co., New Orleans, LA; Glen Williams, Envision Studio, Harrisburg, Pennsylvania; Dennis Winiarski, Knutson Brothers, West Allis, WI; Jean Carmichael, NAHB Software Review Program (Chapter 5); Allan Freedman, Builder Business Services, NAHB; and David Jaffe, Office of Staff General Counsel/Legal Services, NAHB (Chapter 7).

This book was produced under the general direction of Kent Colton, NAHB Executive Vice President, in association with NAHB staff members Jim DeLizia, Staff Vice President, Member and Association Relations; Adrienne Ash, Assistant Staff Vice President, Publishing Services; Rosanne O'Connor, Director of Publications; Sharon Lamberton, Assistant Director of Publications and Project Editor; David Rhodes, Art Director; John Tuttle, Publications Editor; and Carolyn Poindexter, Editorial Assistant.

Introduction

In this book, we explain current techniques for scheduling several residential construction projects. We will explore manual and computer methods, communication of the schedule, and legal issues. We will schedule construction for a freestanding garage, a house, and a remodeling project as examples.

To be a successful builder or remodeler, you must organize, staff, coordinate, monitor, and control what is taking place on one or more jobsites. Formal planning and scheduling techniques have been developed to help perform these operations effectively so that construction projects can be completed on time and within budget.

The terms *planning* and *scheduling* mean the same thing to many people. However, they are really separate operations. *Planning* has to do with setting goals and priorities and organizing activities. *Scheduling* is the creation of a time-based, detailed representation—usually graphic—of the plan.

Graphic scheduling techniques have been used during most of this century. Early in the 1900s, Henry Gantt developed charts used to organize activities. Known today as Gantt charts or bar charts, these scheduling devices are widely used and easily understood. We will explore the uses and limitations of these charts in Chapter 2.

The basis of what is now referred to as the critical path method (CPM) of scheduling was developed by the U.S. Navy and DuPont. Original CPM concepts were used to schedule renovations to refineries and the construction of nuclear submarines. These original concepts have been customized to apply to the scheduling techniques we call CPM.

Initial construction schedules used the CPM with the *activity-on-arrow* (AOA) or *arrow diagram* method. Subsequently, the *precedence diagram method* (PDM) emerged as an alternative methodology. PDM plots activities using diagrams with nodes that represent the activity. This *activity-on-node* (AON) method of drawing gives planners a more convenient way of planning, sequencing, and graphing the plan.

Today, CPM diagrams are used to represent almost any planned event and the information generated can be extremely useful. A good schedule can help identify potential time overruns, cost overruns, and conflicts with other projects; coordinate complex operations; and make the best use of the only non-renewable resource available to the builder or remodeler—time.

When will each of the major activities take place when building a typical house? For a project similar to the plans shown in Appendix A, many builders might be able to guess with a fair degree of accuracy. But can you afford to guess when there are ways to predict accurately? Just as a good set of prints acts as a guide for building the house, a well-developed schedule acts as a guide for organizing and controlling the construction process. Today's builders must use the best managerial techniques available, including scheduling, to continue to survive and prosper.

An integrated planning and scheduling system offers you:

- Superior organization and control throughout the project
- Time savings during the construction of each project
- Reduced overhead rates, since more units can be constructed using the same staff
- Reduced construction interest
- Improved relations with subcontractors, material suppliers, and financial institutions
- Improved customer satisfaction with more consistent on-schedule project delivery

Today, many builders or remodelers use schedules to:

- Organize their projects and "build" them on paper
- Coordinate their subcontractors, suppliers, and owners and tell them in advance when they are expected to perform
- Determine "overload" of resources
- Recognize and correct delays in the project before they cause major problems

Too often, builders and remodelers find themselves managing their businesses by reacting to one crisis after another, "putting out fires." But if you consistently use the scheduling techniques as outlined in this book, you will find that you can control your work instead of the work controlling you. The methods presented here allow you to plan and schedule to any level of detail needed.

Even if your company has a scheduling department, anyone who uses a schedule should be trained in its use to make the most of what it offers.

Scheduling can be done either manually or by using increasingly inexpensive computer hardware and software. With time of the essence, manual scheduling requires more time than many builders are willing to expend. Computers allow for quick changes and updates, copying existing schedules, and automatic performance of many advanced operations. However, an up-front investment of time, money, and training is required if you want to fully utilize a computerized scheduling system.

Builders and remodelers will naturally develop their own techniques for putting the schedule on paper and communicating with subcontractors, owners, and other interested parties. Each business will develop a method that suits its particular environment, accommodating factors such as whether you employ a crew or work wholly with subcontractors, the size of your office staff, whether the office has up-to-date computer equipment, whether projects are centralized or on scattered sites, and the volume of work. Smaller builders may be in the habit of simply writing their schedules out on a calendar, whereas larger builders may find that the most sophisticated computer program doesn't completely meet their needs. The bottom line is that whatever method is used, the schedule is a critical management tool for planning the job and communicating job progress. Whether you are working on paper or on computer, following the principles outlined in this book will help you improve your project schedules.

Chapter 1 presents steps that should be taken in planning the project: organizing different phrases of the work and developing a detailed breakdown of all the activities that must take place. Chapter 2 introduces various techniques for diagramming the activities in a schedule, including bar charts and more sophisticated methods.

Chapter 3 shows the development of a sample plan and schedule for construction of a house. Of course, seldom does a project go exactly as initially planned and scheduled. The builder or remodeler needs to periodically review the schedule, identify existing problems, and make corrections to keep the project on track. Chapter 4 discusses the monitoring of schedules. Chapter 5 addresses the use of computers in scheduling. Chapter 6 explains the development of a detailed schedule for a remodeling project, and Chapter 7 presents information on the legal implications of scheduling. Plans and cost estimates for the sample house and sample remodeling project appear in appendixes at the back of the book.

Chapter 1

Planning for the Schedule

A plan is a way of making or doing something that has been worked out beforehand. Builders follow *plans* for constructing a house or remodeling a kitchen. They also follow *specifications,* which further define the materials and services required for the project. These construction documents provide you with a predefined set of guidelines that describe the completed project.

You must arrange to purchase the materials, labor, and buy or rent the equipment necessary to complete each project. What should be done first? When will different materials be needed on the site? How long will it take to finish the project? To answer these questions, you create a construction plan.

To create the plan, you need to review the construction documents and determine what tasks are to be done, how long each task will take, and in what order the tasks will occur. Organizing this information in the plan helps you to estimate when materials will be needed, what should happen each week, and when the project is planned to be finished.

Looking back, anyone can give the right answer, the right bid, or the right completion date. However, when you create a construction plan and schedule, you must anticipate events and circumstances that will happen in the future. Some builders and remodelers simply guess, based on past knowledge, how long a project will take. They mentally assess the size of the project, time of year the project will start, people to be employed, material availability, and several hundred other factors to determine the completion date. While this method works for many builders, there are more scientific ways to schedule a construction project. Systematic planning and scheduling can allow you to develop a consistent way of organizing your work and providing information about each project to customers, suppliers, subcontractors, and employees.

FORMAL PLANNING TECHNIQUES

Implementing a planning and scheduling system is a fairly simple process. You must commit to follow an organized series of steps that should be established, maintained, and used. Most importantly, the system should be developed and tailored to meet the specific needs of your business to enable you to complete your projects.

The scheduling system should be written down so that it can be consistently communicated to employees and subcontractors, customers, and other parties such as financial institutions, real estate agents, and developers.

A good scheduling system contains three key elements: planning, scheduling, and monitoring. Figure 1.1 shows the flow of information among these elements.

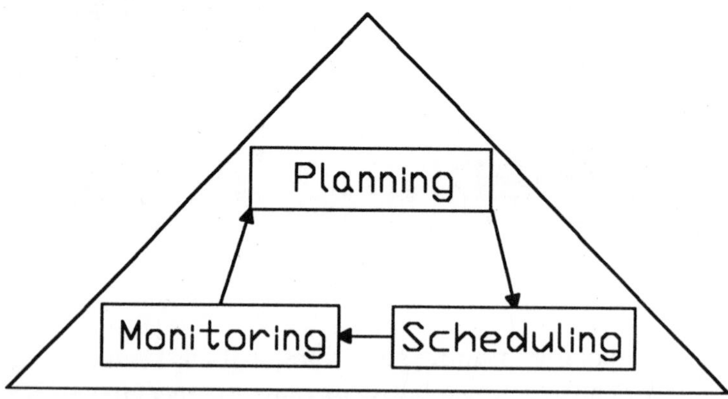

Figure 1.1: Company Scheduling System

Information first flows from the planning phase to the scheduling phase. The schedule is then monitored. (Monitoring can be as simple as visual inspections or as formal as written reports.) Monitoring uncovers any need to change the plan in the event of problems, changes in circumstance, or poor initial planning. This information then initiates a new cycle of planning, scheduling, and monitoring. Followed consistently, a good scheduling system makes it easy for you to check project status, correct errors, and adjust for contingencies.

Planning is the most important phase. In this phase, you first define and organize the work to be accomplished. The most common way of defining the scope of work is by creating a list of tasks or activities. Then you organize the tasks by determining the relationships between each task.

Builders who build the same or similar projects often can work from a master task list with minor variations. Remodelers usually deal with custom designed projects and may have to completely review every project to develop the tasks. Different builders and remodelers will come up with different lists of tasks and different sequences for the same project. Because projects—and builders and remodelers—vary, the sequence of activities will vary for each project.

Some builders stop using their scheduling system once their initial planning is complete. They don't see any additional benefits in creating a schedule. Having identified and organized the tasks, they want to move on to construct the project. Although some benefits are realized by simply identifying and organizing the tasks, the next phase—scheduling—will prove even more worthwhile. Scheduling sets time limits, predicts when parts of the project will be completed, and helps to verify the accuracy of the original plan.

Monitoring the project is the most neglected phase in the scheduling process. A properly monitored schedule enables you to minimize delays, forecast delays, and resolve conflicts at the earliest opportunity. Yet many people find it psychologically easier to "fight fires" than to exercise the discipline required to gain control over their project schedules.

DEFINING AND ORGANIZING ACTIVITIES

The first step in planning is to identify the activities or tasks to be accomplished. Begin by creating a standard list of the *phases* of the project. A phase can be defined as a title or heading for different groupings of activities that you may wish to use in organizing the project. One phase listing could use the stages that a project goes through from start to finish. The phases will be used to organize the project to minimize omissions and confusion. Since the phases are organizational headings, each builder can develop a unique set of phases. Eventually, the list becomes a checklist for all similar projects.

The following is an initial list of phases that a builder might create:

- Project Startup
- Initial Sitework
- Foundation Phase
- Slab-on-Grade Work
- Structure Phase
- Structure Dry-in Phase
- Finish Phase
  - Custom Items
  - Finish Sitework
- Project Close-out

A remodeler might create the following list:

- Project Startup
- Protect Existing Structure
- Demolition Work
- Interior Kitchen Rework
- Interior Bath Rework
- New Construction Work
- Project Close-out

It helps to apply the same list in planning several projects before changing it. Using a standard set of phases allows you to collect and compare information over a period of time. Using a different organization makes it difficult to compare results.

The next step is to break down each phase in the list into a more detailed list of activities. Each part of the list can be subdivided to show as much detail as you require.

One easy way to organize the details on paper is to use indents to show how the activities fit within each phase. For example, activities in the sitework phase might break down as follows:

Initial Sitework
- Clear Site
  - Cut Down Marked Trees
  - Haul Off Trees
  - Remove Stumps
  - Remove Vegetation and Topsoil
- Grade Building Area
  - Haul in Dirt
  - Haul Off Dirt

Compact Dirt
Site Concrete Work
Grade Driveway and Walks
Install Concrete Driveway
Form and Place Sidewalks
Site Landscaping
Spread Topsoil and Fine Grade
Install Plants at Building
Place Sod
Place Seed and Cover

In the list above, some activities are presented in greater detail than others. How far to break down the activities is a matter of individual preference. Some projects will require more detail, others less. You do not have to include all the detailed tasks for every project, but the schedule should be detailed enough to allow you to meaningfully monitor and coordinate the work.

You can now create a master list that contains a breakdown of all the phases you plan to use. It is critical to list every possible activity so that you will not overlook anything important when planning your schedules. Many items in your detailed list will seem extremely obvious, such as the activities listed under Clear Site and Grade Building Area. However, when you are working with a new subcontractor or with someone who has had difficulty performing to your expectations, the detail will help you communicate more clearly and completely.

Using the master list as a guide, you will select the activities that meet the needs of a particular project. The key is to provide enough detail to manage the work without becoming bogged down. The following tips will help you determine the level of detail needed in your project list:

- Separate activities by subcontractor or individual. This will reduce confusion about who is responsible for accomplishing which activity.
- Assess lead times for ordering and receiving materials for each activity. Materials with long lead delivery times need to be tracked. For example, if the project requires an Italian marble floor with a ten-week delivery time, it is essential to have an activity for ordering the material.
- Separate similar activities that occur at different times. For example, if electricians need to be on the project with different materials at different times for rough-in and trim-out, two separate activities should be used.
- Use the project estimate as a guide. If a construction step is required in the estimate, it should be included in an activity in the schedule. Attempting to track every item listed in the estimate in detail, however, can take more time than it's worth. Activities with small durations and/or cost impacts can be combined.
- Talk with subcontractors, owners, lenders, and building inspectors, and ask them to identify activities they believe are important. These people understand their areas of work and can help you avoid costly delays. Many builders are uncomfortable with providing detailed schedules to anyone outside the company. If you do provide CPM schedules to subcontractors, consider sharing only the early start dates.

LISTING ACTIVITIES

Three examples are used in this book to demonstrate how to plan and schedule a project. The first example is a simple freestanding garage addition requiring relatively few tasks. The second example is a complete house schedule based on the plans and estimate that appear in Appendixes A and B. The third example is the remodeling of a kitchen in the same house. The plans and estimate for the kitchen are located in Appendixes C and D. The master list of activities that appears in Figure 1.2 will be used in developing the schedule for the house. Remember, this list is not in a construction sequence. This list simply identifies the activities that we will use for the project to be scheduled. The schedule itself will place the activities into the proper sequence.

Figure 1.2: Master List of Activities for Constructing a House

- Project Startup
  - Obtain Permit
  - Obtain Approved Architecture Drawings
  - Obtain Plot of Site
  - Obtain Temporary Power
  - Obtain Temporary Water
  - Obtain Builders Risk Insurance
  - Surveyor to Locate Site
  - Surveyor to Locate Building Corners
- Owner Activities
  - Owner Select Plumbing Fixtures
  - Owner Select Light Fixtures
  - Owner Select HVAC System
  - Owner Select Paint
  - Owner Select Cabinets
- Initial Sitework
  - Site Grading
    - Clear Site
    - Grade Site
- Foundation Phase
  - Install Footings
    - Set Batter Boards
    - Dig Footings
    - Set Grade Stakes
    - Footing Inspection
    - Pour Footings
  - Install Foundation Block
    - Order Block
    - Deliver Block
    - Order Sand and Mortar
    - Deliver Sand and Mortar
    - Lay Foundation Block
- Slab-on-Grade Work
  - Utility Rough-in
    - Plumbing Rough-in
    - HVAC Rough-in
    - Electrical Rough-in
    - Radon System Rough-in
    - Dryer Duct Rough-in
  - Slab Preparation Work
    - Fine Grade Subbase
    - Lay Out and Set Batter Boards
    - Excavate Turndown Edge
    - Set Edge Forms
    - Spread Gravel
    - Termite Treatment
    - Place Poly and Welded Wire Mesh
    - Set Grade Stakes
  - Place Slab
    - Inspect Slab
    - Place and Finish Concrete
    - Cure Concrete
- Structure Phase
  - Erect Rough Framing
    - Order Framing Material
    - Deliver Framing Material
    - Floor Framing and Decking
    - Walls and Sheathing
    - Deliver Roof Trusses and Roof Framing
    - Roof Framing and Decking
  - Complete Rough Framing
    - Check Plumb and Square
    - Check Door Openings (size and square)
    - Install Deadwood and Blocking
    - Wall Insulation
    - Framing Inspection
    - Clean Up Waste Material
  - Uitility Rough-in
    - Deliver Tubs and Showers
    - Plumbing Top-out

(continued)

Figure 1.2: Master List of Activities for Constructing a House (continued)

- Plumbing Inspection
- HVAC Rough-in
- Electrical Rough-in
- Electrical Inspection

Structure Dry-in Phase
- Exterior Wood Items
  - Install Fascia and Soffits
  - Order Exterior Doors and Windows
  - Deliver Exterior Doors and Windows
  - Install Exterior Doors and Windows
  - Install Siding
- Exterior Masonry Items
  - Order Brick
  - Deliver Brick
  - Install Brick
- Roofing Items
  - Install Flashing and Felt
  - Install Shingles

Finish Phase
- Interior Wall Material
  - Order Paneling
  - Deliver Paneling
  - Install Paneling
  - Order Sheetrock
  - Deliver Sheetrock
  - Hang Sheetrock
  - Tape and Finish Sheetrock
  - Clean Up Sheetrock Waste
- Wall Finish Items
  - Paint Interior Walls
  - Paint Exterior Walls
  - Deliver Wallpaper
  - Hang Wallpaper
  - Paint or Stain Interior Trim
  - Paint or Stain Interior Doors
  - Paint Touch-up
- Interior Wood Trim
  - Deliver Interior Trim
  - Install Interior Trim
    - Base Material
    - Chair Rail Material
    - Crown Molding Material
  - Install Closet Shelving and Rods
  - Install Interior Doors
- Finish Flooring
  - Install Ceramic Tile
  - Install Other Hard Tile
  - Install Vinyl Tile
  - Install Carpet
  - Install Wood Flooring
    - Place Wood Flooring
    - Sand and Finish Wood Flooring
- Install Miscellaneous Items
  - Insulate Attic
  - Order Cabinets
  - Deliver Cabinets
  - Install Cabinets
  - Install Bathroom Accessories
  - Order Appliances
  - Deliver and Set Appliances
    - Dishwasher
    - Stove
    - Compactor
    - Washer/Dryer
    - Range Hood
- Finish Utilities
  - Plumbing Trim-out
  - HVAC Trim-out
    - Deliver HVAC Equipment
    - Set Grills and Thermostat
    - Start Up and Test Equipment
  - Electrical Trim-out
    - Install Fixtures, Devices
    - Order Light Fixtures
    - Deliver Light Fixtures
    - Hang Light Fixtures
  - Hook Up Main Power
  - Check and Test System
  - Final Electrical Inspection
- Custom Items
  - Install Fireplace Items
    - Order Fireplace
    - Deliver Fireplace
    - Set Fire Box
    - Set Flue and Box-in
  - Install Outside Decks
  - Install Hot Tub
    - Order Hot Tub
    - Deliver Hot Tub
    - Set Hot Tub
  - Install Alarm System
  - Install Telephone Wiring
  - Install TV Wiring
- Finish Sitework
  - Grade Driveway and Walks
  - Form and Install Concrete Driveway
  - Form and Place Sidewalks
  - Site Landscaping

Project Close-out Phase
- Remove Temp. Water Connection
- Remove Temp. Power Connection
- Clean House
- Final Inspection
- Complete Punch List Work

The detailed breakdown in Figure 1.2 outlines most of the activities builders can use to plan the construction of a typical house. Other projects may require more and different activities. Some builders will organize the activities in different ways. Some builders will add more detailed information to this list, such as Call Framing Sub, Owner Lay Out TV Outlets, and Verify Electrical Rough-in. This list will reflect how the builder organizes and constructs each project.

ACTIVITY DURATION

Determining the duration of various activities depends as much on builder intuition as it does on science. A builder might know that on a good week, with everyone healthy, sunny 75° weather, and all materials onsite, the crew can frame an average house in five days. Seldom do all the pieces fall into place this neatly. The science comes in determining how much work the crew can do on an average (not perfect) project. The information you need to determine this is derived from your company's past history.

Figure 1.3 shows a sample calculation for a framing activity. Assume that your crew has been able to frame their last five houses in five to seven days. You could compile the table shown in Figure 1.3 using information from your records and work assignment sheets.

The next house you plan to build has 3,100 square feet and is similar to the houses above. How much time should be estimated for the framing crew?

One method of making the calculation is to make a guess based on the average of the production rates calculated as follows: (500+590+600+568+725) ÷ 5 = 596.6 square feet each day. Rounding the number to 597, this gives the answer of 3,100 square feet ÷ (596 square feet each day) = 5.2 or 5½ days. Another method to check productivity would be to use the numbers shown in the "Total" line in the chart, yielding an answer of [3,100 square feet ÷ (590 square feet each day)] = 5.25 or 5½ days. Rounding to the half day gives the same duration. This estimate of the duration for the activity fits within the pattern of your table, and probably would give the crew adequate time to complete the task.

Figure 1.3: Sample Duration Calculation

Project	Size	Date Start	Date End	Duration	Production
House 1	2,500 sq. ft.	July 29, 1994	Aug. 4, 1994	5 days	500 sq. ft./day
House 2	3,250 sq. ft.	Aug. 15, 1994	Aug. 22, 1994	5½ days	590 sq. ft./day
House 3	3,000 sq. ft.	Aug. 8, 1994	Aug. 12, 1994	5 days	600 sq. ft./day
House 4	3,980 sq. ft.	July 5, 1994	July 13, 1994	7 days	568 sq. ft./day
House 5	2,900 sq. ft.	June 28, 1994	July 1, 1994	4 days	590 sq. ft./day
Total	15,630 sq. ft.	—	—	26½ days	590 sq. ft./ day

However, some builders prefer to use one of the historical extremes for determining the duration:

Optimistic builder: 3,100 square feet (725 square feet each day) = 4.3, or 4½ days.
Pessimistic builder: 3,100 square feet (500 square feet each day) = 6.2, or 6½ days.

Any of these answers could be considered correct. Many factors can influence the duration of an activity, and not all of these factors can be predicted, so it is helpful to think of the duration as an approximate rather than an exact target.

The following tips will help you determine durations for various activities:

- Remember, historical rates of production should be treated as a guide, not as an absolute. Because of varying circumstances, the same crew may perform exceptionally well on one house but show poor production on another.
- Round the duration *up* by a half a day to allow for equipment set up, tear down, and so forth. Shorter time periods can be scheduled, but most short-duration activities will take a half day to complete. Setting grade stakes may only take an hour, for example, but by the time the instruments are set, the material set out, and the operation completed, a half day is probably gone.
- Have the people doing the work help you determine the duration. The people doing the work must commit the time allotted for doing the task. If they think the duration is feasible, this will increase the chance of things happening on schedule. If you unilaterally set the time frames, the workers' performance may vary, depending on how they view the time given to do the work. If they believe the time is too short, overruns are likely because they won't be motivated to meet an "unrealistic" schedule. Conversely, if too much time is given, the workers may slow down to fill the time. Both extremes cause problems. Use communication and teamwork to your advantage and get people working toward the same goal—completing the project on time.

ADDITIONAL ACTIVITY INFORMATION

Once the activities for a project have been identified from the master list, a *project list* is created. After you take a look at additional information that may affect the job schedule, these activities are assigned estimated durations. Assignment of responsibility, estimated cost, earned value, and resources (crews, material, and equipment) can help make the schedule a more flexible and useful management tool.

Responsibility

For each activity, you must identify who is doing the work. For each activity, one subcontractor or individual should be responsible for completing the task, and this information should be noted on the project list.

When the projects are scheduled, you can quickly see who is doing what and, if multiple projects are being scheduled, identify potential conflicts. Having the responsibility assignments clearly marked can help communication when this information is shared with workers or subs.

Estimated Cost

The estimated cost of each activity is the amount the builder expects to have paid when the activity is finished. The figure may reflect estimates given by subcontractors, an estimated amount for labor, material, and equipment based on a formula, or a combination of both types of estimates.

By appending estimated costs to the activities in the schedule, you can determine amounts owed to subcontractors based on work completed and can anticipate the cash demands that will arise during the project. This information also can be used to compare the costs of specific activities with costs for similar projects to determine if any costs are out of line. Many builders do not track costs according to the schedule. If you have a scheduling system in place that performs well, however, this can be a convenient way to organize cost information.

Earned Value

Earned value is the amount of money a customer or owner will "pay" the builder or remodeler for each activity. The amount generally equals the cost of the activity plus an appropriate amount of the builder's overhead and profit. The total earned value for all activities equals the amount of the builder's contract.

Billing is often done by a *schedule of values* established at the beginning of the project or by some milestone agreement, such as payment of one third when the slab is placed, and so forth. Establishing earned values allows the builder to use the schedule to monitor and anticipate payments from the owner or cash withdrawals from a construction loan. As with costs, using the schedule to track earned values will only be accurate if the builder's scheduling system is running smoothly.

Resource Assignments

Resource assignments identify the crews, material, and equipment needed to complete the activity. When tied to the schedule, this information gives you a picture of how much of any resource is needed on a project at any point in time. This allows for flexible management of the resources so they are used to the fullest extent. This information can also be helpful when juggling multiple projects requiring the same resource.

An example is a builder who has one backhoe. This piece of equipment digs the foundations, helps with fine grading and cleanup, and serves as a lift for framing material. If the builder has three jobs going on, this will be a busy piece of construction equipment. Where

Figure 1.4: Typical Activities With Responsibilities Assigned

Activity	Duration	Responsibility	Resource	Estimated Cost	Earned Value
Wall Framing	1½ days	Framing Sub	3 Carpenters	$2,549	$3,000
Check Framing	½ day	Builder	1 Carpenter	$150	$200
Blocking & Deadwood	½ day	Framing Sub	1 Carpenter	$150	$200

will delays occur? What happens if the machine breaks down for a week? If scheduling conflicts or problems can be identified and eliminated in the planning stage, their impact on the schedule will be reduced.

Figure 1.4 presents several activities that may appear on a project. The activities have the additional items discussed above assigned to them.

SUMMARY POINTS AND TIPS

- A construction plan and schedule represent the builder's or remodeler's attempt to organize activities and anticipate future events.
- Systematic planning and scheduling techniques provide builders and remodelers with a guide that can be used, adjusted, and reused.
- The first step in planning is to create a framework to organize the activities required for any project. Using a standard list—like our list of phases—as the basis for all similar projects allows for project comparisons and helps prevent omissions.
- Builders and remodelers can adjust the level of detail at which they track items on the list.
- A master list should be developed that contains all activities organized by phases or sequence of construction. Be sure to include activities for inspections, owner selections, ordering of materials, or any other applicable items.
- Project-specific lists can then be developed from the master list. Project lists identify the activities that are planned for the specific project.
- Projections of activity durations are based on a combination of past experience, feedback from vendors and subcontractors, and intuition. In making projections, use historical production figures as a guide, not as an absolute.
- Additional information can be assigned to activities on the list to make the schedule more useful.

Chapter 2

Methods for Scheduling Your Plan

In this chapter, we will develop a plan to construct a freestanding garage. We will apply the topics (responsibility, resources, and cash flow) discussed in the previous chapter to this small project. In real life, it is unlikely that a builder or remodeler would go to such lengths for such a small project. However, using a small example will allow us to discuss how the principles apply without having to review a lot of activities.

Many builders have been successful by communicating scheduling information orally. Deliveries are coordinated by making telephone calls at 4:00 p.m. to request materials for the next day. Equipment arrives at the site when needed and people continue working efficiently. However, the techniques discussed here may improve what you are doing and may allow you to get more work done before 4:00 p.m.

One company started having all their project supervisors schedule by creating *logic diagrams* (which will be discussed later in this chapter). After initial training and direction in scheduling principles, they began. Four months later, the company determined that their project durations had been reduced by one week. They also discovered that the supervisors were more confident in projecting estimated completion dates.

When developing an initial plan on which the project schedule will be based, the builder or superintendent must study the plans, the specifications, and the estimates and determine the overall scope of the work to be completed. It is essential that the person in charge of the project help with the planning phase of the scheduling process.

Once the necessary activities have been identified, the builder can develop the logic for the plan and create a graphic chart of the schedule. *Bar charts* and *logic diagrams* are two common ways of graphically presenting the schedule. A bar chart simply lists the activities in relation to a calendar. Reference points on the chart mark when each activity is expected to be accomplished. A logic diagram presents additional information about how various activities interrelate.

GANTT OR BAR CHARTS

The bar chart is a familiar technique for displaying schedule information. Deadlines and durations pertaining to equipment, workers, and materials are easily displayed and quickly understood for use at all levels of management.

A bar chart is useful for communicating with workers and subcontractors. By using a bar chart, everyone involved in the project can quickly see when their activities need to be completed. Bar charts are convenient for the builder or remodeler who is trying to delegate resources efficiently on a tight schedule. Bar charts also are helpful for subcontractors, who are trying to complete their work as quickly as possible.

Creating a Bar Chart

The first step is to identify the activities from our Master List that was developed in Chapter 1. Figure 2.1 presents a table of activities we have selected from the list for constructing a freestanding garage with a slab-on-grade foundation and an unfinished interior. The activities in the table will be used to create a bar chart. (We will create a more detailed schedule for construction of a sample house in a later chapter.)

Figure 2.1: Activities for Constructing a Garage

Description	Duration	Responsibility	Earned Value
Slab Prep Work	4 days	Concrete Crew	$4,000
Inspect Slab Prep	1 day	Inspector	$0
Place & Finish Slab	1 day	Concrete Crew	$1,500
Erect Wood Framing	5 days	Framing Crew	$2,000
Complete Rough Framing	2 days	Framing Crew	$500
Install Shingles	2 days	Roofing Crew	$1,500
Install Door & Windows	1 day	Framing Sub	$1,000
Install Siding & Cornice	2 days	Siding Sub	$2,500
Rough-in & Trim Electrical	2 days	Electrical Sub	$1,000
Paint Exterior	3 days	Painting Sub	$1,000
Final Inspection	1 day	Inspector	$0

As detailed in the table, this project will have eleven activities involving seven different parties and an estimated value of $15,000. We will assume that the builder has discussed this list with the owner and all subcontractors and checked the subcontractors' project estimates when planning the schedule and assigning durations and earned values. (*Earned Values* are dollar amounts assigned to activities that equal the cost to the owner of that activity. Earned Values include all the builder's or remodeler's direct costs plus a prorated share of overhead and profit.)

In this example, the builder will directly supervise three separate crews: the concrete crew, framing crew, and roofing crew. The balance of the activities will be subcontracted.

Some builders subcontract all the work activities and are construction managers. Other builders maintain their own crews and perform many of the onsite construction activities. The responsibility assignments can be changed to meet each builder's management style.

The earned values assigned to the activities equal the amount the builder will be paid for completing each activity. These are determined by calculating the cost of each activity and adding an appropriate amount of overhead and profit. Assigning this information will allow us to obtain a simple cash-flow chart for the project later in the chapter.

Figure 2.2 shows a bar chart constructed from the information in the table. The activities are listed down the left side of the chart. For each activity, a bar is then drawn representing the days during which the activity will be performed.

Figure 2.2: Sample Bar Chart

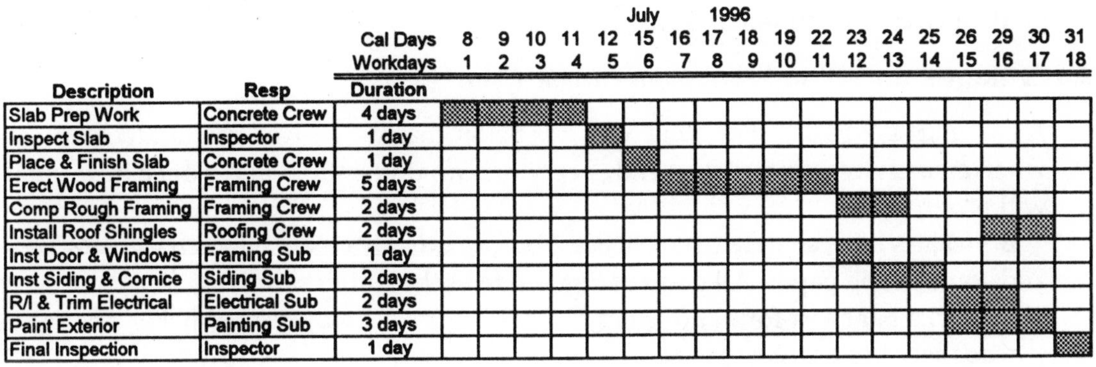

July 1996

		Cal Days	8	9	10	11	12	15	16	17	18	19	22	23	24	25	26	29	30	31
		Workdays	1	2	3	4	5	6	7	8	9	10	11	12	13	14	15	16	17	18
Description	**Resp**	**Duration**																		
Slab Prep Work	Concrete Crew	4 days	■	■	■	■														
Inspect Slab	Inspector	1 day					■													
Place & Finish Slab	Concrete Crew	1 day						■												
Erect Wood Framing	Framing Crew	5 days							■	■	■	■	■							
Comp Rough Framing	Framing Crew	2 days												■	■					
Install Roof Shingles	Roofing Crew	2 days																■	■	
Inst Door & Windows	Framing Sub	1 day												■						
Inst Siding & Cornice	Siding Sub	2 days													■	■				
R/I & Trim Electrical	Electrical Sub	2 days															■	■		
Paint Exterior	Painting Sub	3 days															■	■	■	
Final Inspection	Inspector	1 day																		■

Because activities may overlap, it is impossible to list all the activities in a linear sequence. Activities can be rearranged later if desired. The important items to consider when placing the activities on the chart are: (1) What must be done before this activity can start? and (2) Could a delay in this activity delay the start of other activities?

In our example, it is clear that installation of the slab must be finished before erection of the wood structure may begin. The relationship is represented in the chart by placing the start date for erecting the wood structure at Workday 7, after the slab is completed on Workday 6.

Every activity in the chart is related through the construction process to all the other activities. For the bar chart to be useful, we must be concerned with activities that have a *direct* effect on each other. A direct relationship exists when an activity must be completed before another activity can begin, such as the slab and the framing as shown in the bar chart. A bar chart will not show the relationships between activities; however, by presenting the schedule in a simple graphic format, the chart may help the builder see patterns or problems (for example, if erect wood framing were scheduled to start before finishing of the slab had been completed).

Not all of the activities in the bar chart are directly related. Some activities can start before others are completely finished. Other activities are placed in the chart according to when the parties involved want them to occur.

The following questions can be answered by using the bar chart (Figure 2.2) and the table (Figure 2.1):

- When does the framing material need to be on the project?
  (On or before July 16.)
- Who is responsible for completing electrical work on time?
  (The electrical subcontractor. In other situations, it might be the builder's own crew.)
- When will the project be completed? (On July 31, after 18 workdays, or 24 calendar days; don't forget to account for weekends when converting to calendar days.)

- How much money has the builder earned at the end of July? ($4,000 + $0 + $1,500 + $2,000 + $500 + $1,000 + $2,500 = $11,500.)
- Which activities are the most important?

The last question cannot be fully answered with the information provided. The bar chart presents all activities as if they have the same degree of importance. All activities must be finished for the project to be complete; however, some activities will have a greater impact on the project's completion if they are delayed.

Advantages and Disadvantages of Bar Charts

A bar chart is a basic scheduling tool. It is better than a "To-Do" list and it clearly shows when an activity will start and end. Everyone can understand at a glance how the work will progress.

Bar charts are also easy to construct. You can quickly create a bar chart using a pencil and some graph paper. All good computerized scheduling programs also can present information in a bar chart format (the bar chart in Figure 2.2 was created using a computerized spreadsheet program).

Just as they are easy to create, bar charts are easy to mark to show the current project status. As long as the work is on schedule, you just have to color in the bars as the work progresses.

One disadvantage of bar charts is that they cannot show relationships between various activities or which activities are critical. The relationships between roofing installation, door and window installation, and electrical rough-in, for example, are clear only to the person who constructed the schedule. It is not possible to determine which activities can delay the project and which activities can be delayed without any impact on the project.

Bar charts can quickly become obsolete when significant delays occur. You must then adjust all start times and redraw all the activities affected by a delay. If a project involving 50 activities experiences several delays, whether caused by late delivery of materials, weather, or other work complications, manual redrawing can become a time-consuming process. Some builders save time by leaving space on the bar chart for updating, or by color coding the chart as updates are added. A line in such a chart might appear as in Figure 2.3.

Figure 2.3: Bar Chart Detail

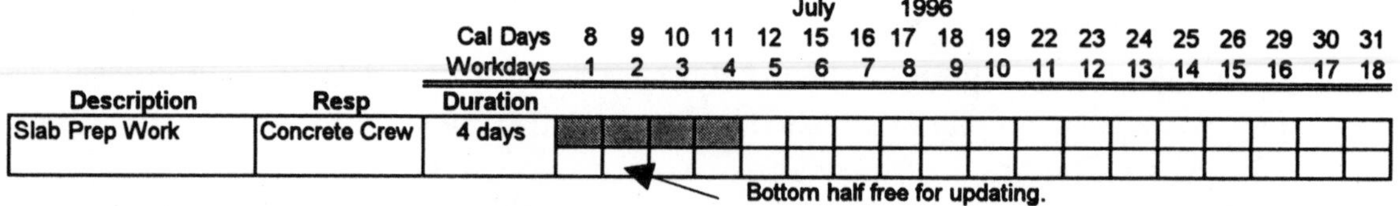

LOGIC DIAGRAMS

You need to perform the same analysis of the project data whether you plan to construct a bar chart or a logic diagram (also called a Pert chart or a critical path method—CPM— diagram). However, a logic diagram can show far more about how the different activities relate to one another. Logic diagrams consist not only of activities but also connectors that show how the activities relate to each other. Calculations are made after creating logic diagrams to determine when activities need to be completed. These calculations are called the *network analysis* or the *forward and backward pass*. If a computer program is used, the calculations are done automatically; however, you should understand the basics of how they are made.

The two most common logic diagrams are the Activity-on-Arrow (AOA) and the Activity-on-Node (AON) diagrams. The AOA method is the older and shows which activity can start when another activity is finished—finish-to-start logic. AON diagrams use boxes as nodes and are easier for most builders to work with. Most scheduling software available today uses AON diagraming. We will construct examples of both types of diagrams and compare the results.

Both AOA and AON diagrams identify a *critical path*. The critical path is the longest path through the project, given the established activity relationship. It is made up of the activities whose durations, when added together, determine when the project can be completed. The critical path is found by doing the network analysis. If a critical path activity is delayed, the project will be delayed.

Activity-on-Arrow (AOA) Diagrams

An AOA diagram uses arrows to represent activities. The name of the activity is written above the arrow. The duration of the activity is usually noted below and at the center of the arrow. A typical activity is shown in Figure 2.4.

Figure 2.4: Simple Activity-on-Arrow Diagram

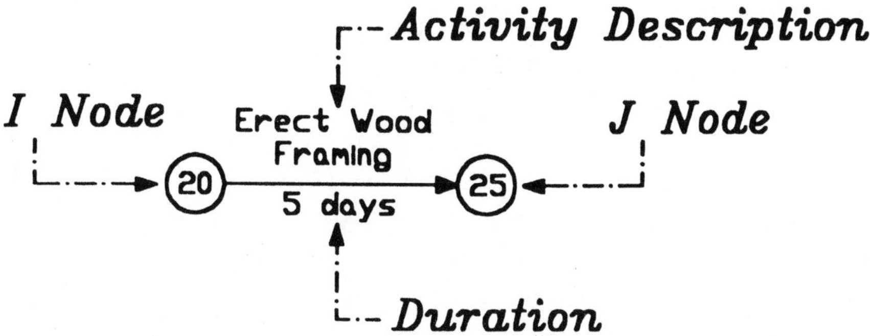

The circles in the diagram represent points in time or *events*. The technical name for one of these circles is *node*. The circle at the start of an activity is the *I* node and the circle at the arrow head is the *J* node.

In Figure 2.4, the I node 20 marks the point in time when wood framing can begin. J node 25 represents the day wood framing will be completed. Any activity that cannot start until the framing is complete must appear on an arrow after node 25.

The numbers in the circles have no special significance other than to help identify that particular activity. Thus, 20-25 identifies the activity Erect Framing. The numbers do not relate to calendar days or workdays. Numbers are used so arrow diagrams can be keyed into a computer program and to help clarify the start and end of the activities. Figure 2.5 shows two activities.

Figure 2.5: Simple AOA Diagram Showing Two Activities

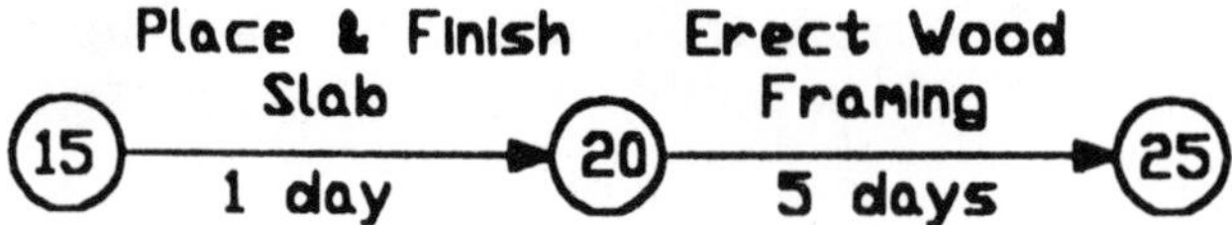

When more than one activity has to be completed before the next activity can begin, all are shown going into the same circle. They all must be completed before any activity leaving that circle can begin. In Figure 2.6 below, erection of the wood structure cannot begin until the slab has been finished and framing material has been delivered.

Another type of activity that appears on AOA diagrams is called a *dummy* or *restraint*. Dummy activities are assigned a duration of 0 days and appear on the diagram as dashed lines.

Figure 2.6: AOA Diagram Showing Multiple Activities

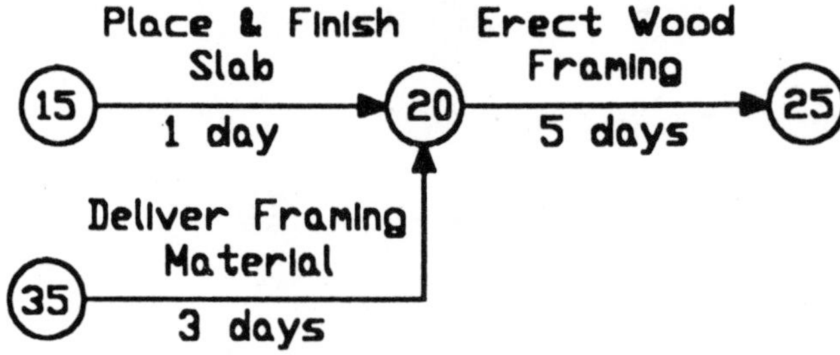

The dummy activity has two uses. It prevents two activities from having identical I and J nodes, which would interfere with the computer's ability to uniquely identify individual activities. When dummy activities appear in the schedule, they do not add additional time to the network calculation. Figure 2.7 shows the right and wrong way to place a dummy to avoid having two activities with the same node numbers. (Note that in the wrong way, two activities have identical I and J numbers.)

The second use of a dummy activity is to help the builder avoid making false relationships between nonrelated activities in the diagrams.

In Figure 2.8, does the delivery of the roofing material have any impact on the electrical rough-in? No. However, because of the way this diagram is constructed, a delay in the roofing delivery could cause a computer-generated schedule to wrongly report that the electrical rough-in cannot be started. Inserting a dummy clarifies the relationship between the activities. As redrawn in Figure 2.9, the electrical rough-in is separated from the node used

Figure 2.7: Placing a Dummy Activity in an AOA Diagram

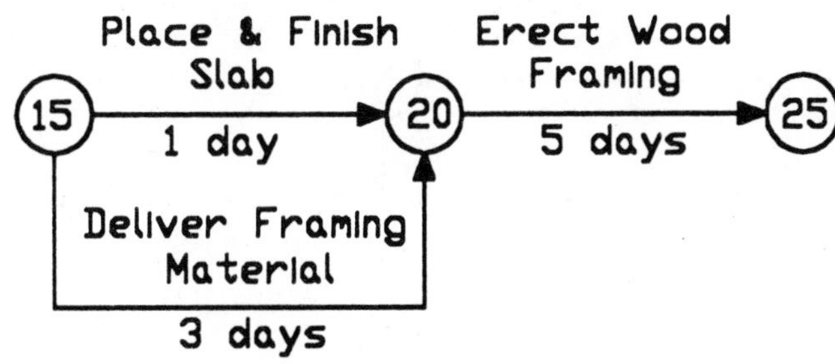

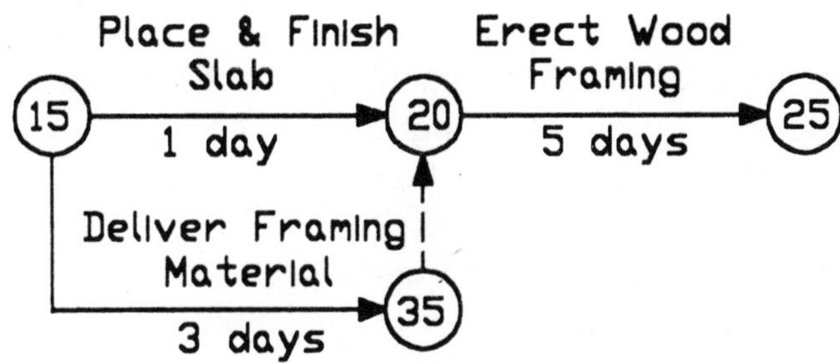

by the delivery of roofing materials. If a delay occurs in delivery of roofing materials, the schedule will no longer show any problem with starting the electrical rough-in.

Developing an AOA Diagram. We will now develop an AOA logic diagram for the garage activities listed in Figure 2.1.

As always, planning priorities need to be identified before creating the diagram. The scheduling priorities for the garage are as follows:

- Complete the project as quickly as possible.
- Make the most efficient use of the crews, equipment and time available.
- Create a working schedule with realistic labor, material, and time requirements.

Figure 2.8: AOA Diagram Showing False Relationship

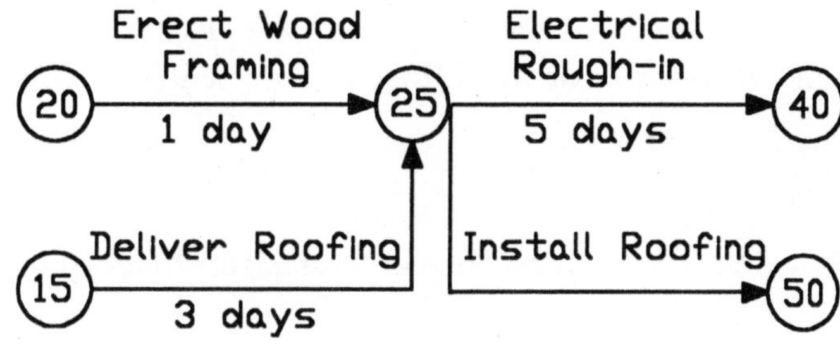

Figure 2.9: AOA Diagram With Dummy for Clarification

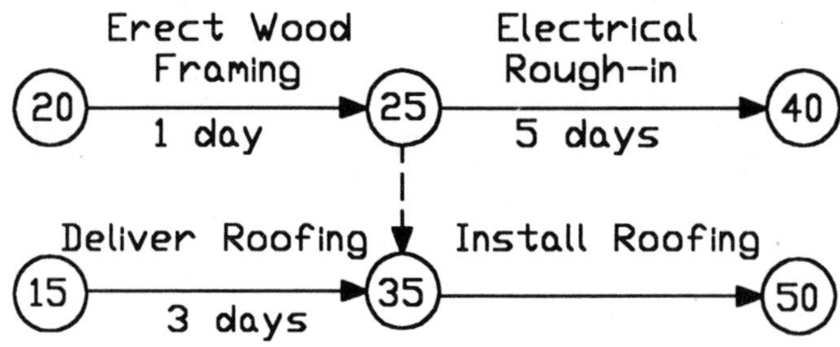

Three kinds of *scheduling restraints* must be accommodated in meeting these priorities: hard, soft, and resource. (Resource restraints are a special type of soft restraint.) A *hard restraint* involves an absolute requirement: Some activities simply cannot start until others are completed. For example, roofing cannot be installed before the roof deck is in place. Therefore, installation of the roof deck acts as a hard restraint for roofing.

A *soft restraint* represents a condition that is not absolutely necessary but that has been incorporated for the convenience of the builder or remodeler. An example of a soft restraint would be that the vinyl floor will not be installed until after the painting is completed. Painting restrains the vinyl flooring, but it does so as a result of the builder's preference or option, not as an absolute necessity.

A *resource restraint* reflects the fact that resources are limited. Builders use resource restraints in a schedule to better manage their resources. The resource may be personnel, equipment, material, or money. A simple example would be two activities that require the use of the same piece of equipment. Unless a second piece of equipment is made available, one activity must precede the other. Resource restraints technically are soft restraints, but they are important enough to merit special mention.

A combination of priorities and restraints comes into play in developing any schedule diagram. Whenever priorities and restraints conflict, the builder or remodeler must decide which factor is the most important. For example, our vinyl tile subcontractor may be planning to use the same crew to lay carpet. If we make the sub wait to lay the tile until after the painting is complete, that crew may be needed to lay carpet at the same time. The restraint between the vinyl tile and painting thus causes a problem with two of our priorities: making efficient use of resources and completing the project as quickly as possible. To solve the problem, we may decide to let the vinyl tile be installed before the painting is completed.

Changing a soft restraint once the schedule is in place may cost the builder or remodeler more money or result in a delay.

Figure 2.10 shows the key relationships and restraints that can be included in a schedule for the garage.

In reviewing the diagram, the builder has made several decisions concerning the logic in the diagram. There are several hard relationships: Inspect Slab has to be completed before Place and Finish Slab and the slab has to be in place for the framing to begin.

Figure 2.10: AOA Diagram for Garage Schedule

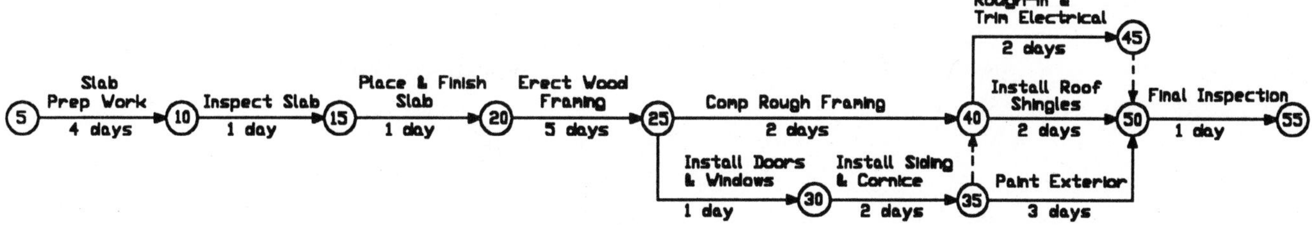

A soft restraint in the diagram is shown by having the Install Siding & Cornice activity delay the start of the Install Shingles and Rough-in and Trim Electrical activities. This is created by adding the dummy from node 35 to node 40. Some builders would feel this is unnecessary and would omit the dummy. Remember, soft restraints may exist that do not appear in the schedule as dummy activities.

The activity Rough-in & Trim Electrical could begin after Erect Wood Framing. We have decided to delay the start of the electrical work until after the framer is finished with the siding and cornice. This avoids having different crews working in the same place at the same time. Also, as the electrician is doing both the rough-in and trim work in the same activity, the exterior siding and cornice must be finished for the exterior fixtures to be placed.

One last item to notice in the diagram is the dummy activity at node 45 to node 50. This dummy has been inserted to avoid having two activities with the same node numbers. The logic would not be affected if the dummy were omitted; however, the proper procedure requires the dummy.

Network Analysis in AOA Diagrams. Once the project plan has been drawn as a logic diagram, we can perform the network calculation or analysis. This process involves four steps:

- Establishing *start* and *finish times* for each activity
- Identifying the *float or slack* available to each activity
- Determining the *critical path*
- Calculating the *overall duration* of the project

The first part of the calculation is called the *forward pass*. In the forward pass, you simply start at the beginning and add the durations to the early start times at the I node of each activity. Where more than one activity ends at the same J node, the *largest* of the two durations determines the early start time for activities following that node. This process is used until all activities have been calculated.

The forward pass provides *early start* and *early finish* times for each activity in terms of workdays. The project begins at the end of Workday 0 and will finish at the end of the largest workday calculated. The workdays can be converted to calendar days once the starting date and all nonworking days are identified for the project. Figure 2.11 shows the process of calculating the forward pass. Slab preparation work begins at 0 and is projected to take four days, for an early finish time (EFT) of 4. The EFT for slab preparation equals the early start time (EST) for the inspection of the slab. Inspection is projected to take one day, resulting in an EFT of 5. The calculation continues in this manner until the end of the job. Dummy restraints have been included in Figure 2.11 to show how dependent and independent activities converge during the forward pass. At node 40, for example, the EST used to calculate rough-in trim electrical work is 13, whereas the EST used to calculate installation of shingles is 14. The EFT for Final Inspection reflects the sums of the largest of all preceding calculations.

The second part of the calculation is called the *backward pass*. To conduct a backward pass, you begin at the end and work backward toward the start of the project, subtracting the durations from the late start times at each J node. When two activities leave a common node, the *smallest* duration is used to calculate the next event's late start time.

The backward pass provides the late start and late finish times for each activity. The late times provide the builder with the latest time an activity can be started and still have the project completed on schedule. The workdays can be converted to calendar days. Figure 2.12 shows the backward pass process.

Once the forward and backward passes have been completed, we have calculated the information needed to manage the project. Figure 2.13 summarizes the information in a simple report format. The calendar dates are shown below each workday and are equal to the days used in the bar chart in Figure 2.2.

The procedure for converting workdays to calendar days can be a little confusing at first glance. Notice that the first activity begins on day 1 and ends on day 4. (We assume that the end of day 0 is the start of day 1. Activities that have a one-day duration start and end on the same day.)

Workdays appear on the calculation diagrams as end-of-day points. Figure 2.13 adjusts the start dates for each activity to the beginning of the day by adding one day to the start number calculated on the diagram.

Figure 2.13 also shows the float for each activity. *Float* can be defined as the amount of time an activity can be delayed without delaying the project. Float is calculated by subtracting the late start workday number from the early start workday number. Activities whose early and late start workdays are identical have no float and are automatically *critical* activities.

Manually calculating the start and finish dates of activities is a fairly easy process. However, it is time-consuming to do so for projects of any size and scope. Computerized scheduling programs can quickly and effectively convert workdays to calendar days and make adjustments. The computer cannot, however, determine the logic necessary to create the initial diagram.

The builder must create the diagram by determining the proper sequence of activities and establishing realistic durations for activities. The calculations performed by the computer are only as good as the information and logic developed by the builder. Be careful.

The AOA diagram:

- Depicts the activities and their relationships to one another
- Identifies the critical activities after the float has been calculated
- Shows early and late start and finish dates
- Sets an estimated completion date for the project
- Shows the flow of the work in a manner that allows you to easily analyze the planning sequence

Figure 2.11: Forward Pass Calculations

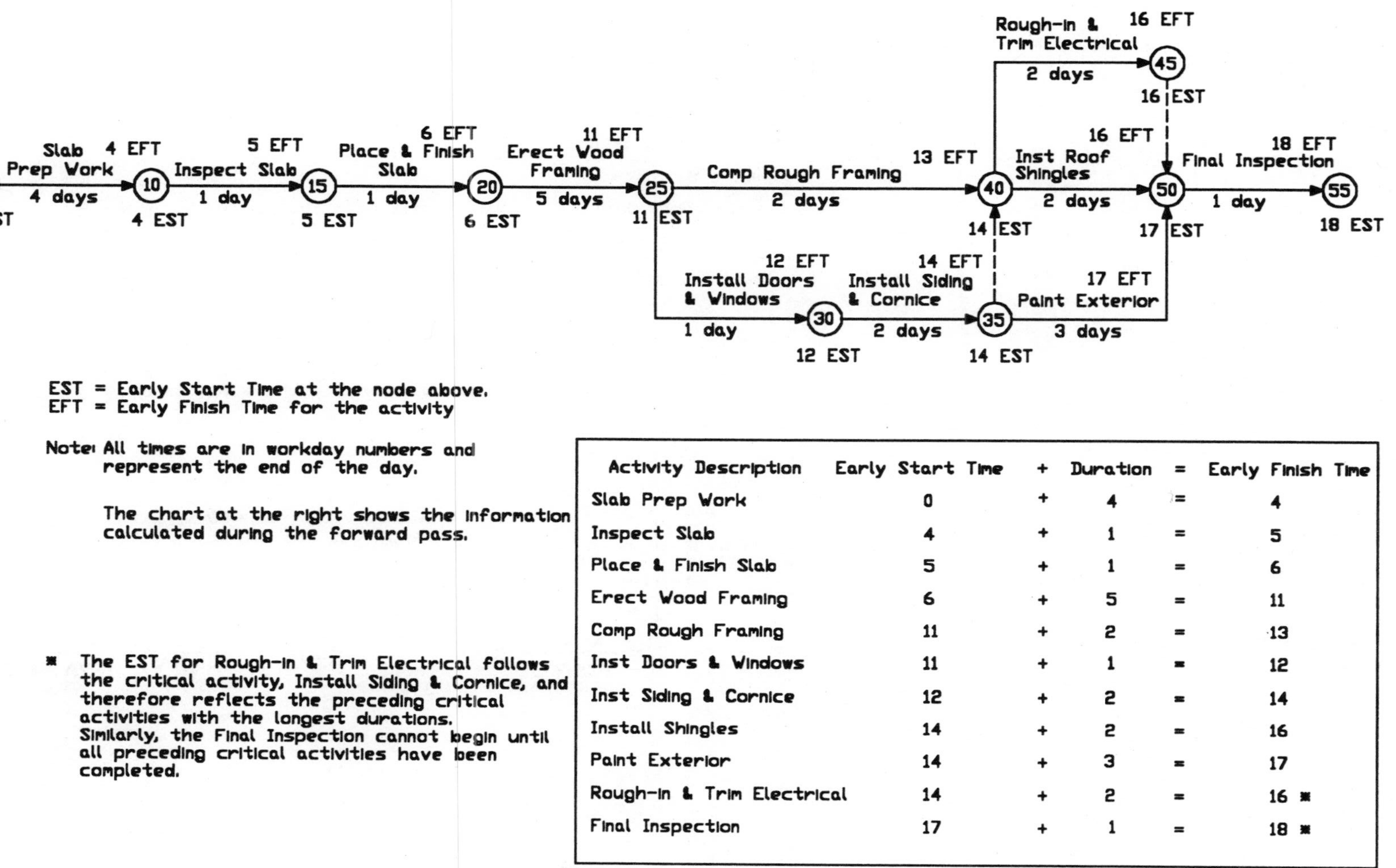

EST = Early Start Time at the node above.
EFT = Early Finish Time for the activity

Note: All times are in workday numbers and represent the end of the day.

The chart at the right shows the information calculated during the forward pass.

* The EST for Rough-in & Trim Electrical follows the critical activity, Install Siding & Cornice, and therefore reflects the preceding critical activities with the longest durations. Similarly, the Final Inspection cannot begin until all preceding critical activities have been completed.

Activity Description	Early Start Time	+	Duration	=	Early Finish Time
Slab Prep Work	0	+	4	=	4
Inspect Slab	4	+	1	=	5
Place & Finish Slab	5	+	1	=	6
Erect Wood Framing	6	+	5	=	11
Comp Rough Framing	11	+	2	=	13
Inst Doors & Windows	11	+	1	=	12
Inst Siding & Cornice	12	+	2	=	14
Install Shingles	14	+	2	=	16
Paint Exterior	14	+	3	=	17
Rough-in & Trim Electrical	14	+	2	=	16 *
Final Inspection	17	+	1	=	18 *

Figure 2.12: Backward Pass Calculations

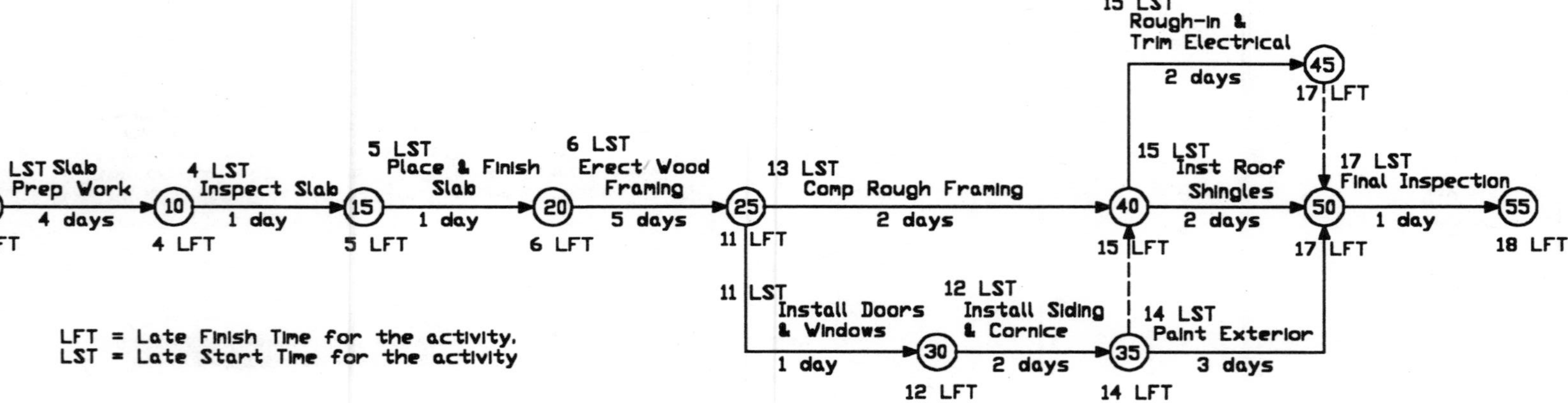

Note: All times are in workday numbers and represent the end of the day.

The chart at the right is the information calculated during the forward pass.

The Late Finish Time of the last activity, Final Inspection, equals its Early Finish Time.

The Late Finish Time of Erect Wood Framing follows the critical path and is therefore equal to the smallest Late Start Time.

Activity Description	Late Finish	-	Duration	=	Late Start
Slab Prep Work	4	-	4	=	0
Inspect Slab	5	-	1	=	4
Place & Finish Slab	6	-	1	=	5
Erect Wood Framing	11	-	5	=	6
Comp Rough Framing	15	-	2	=	13
Inst Doors & Windows	12	-	1	=	11
Inst Siding & Cornice	14	-	2	=	12
Install Roof Shingles	17	-	2	=	15
Paint Exterior	17	-	3	=	14
Rough-in & Trim Elect.	17	-	2	=	15
Final Inspection	18	-	1	=	17

Figure 2.13: Summary Information

Activity	Early Start	Early Finish	Late Start	Late Finish	Float
Slab Prep Work	WD 1 7/8	WD 4 7/11	WD 1 7/8	WD 4 7/11	0 days
Inspect Slab	WD 5 7/12	WD 5 7/12	WD 5 7/12	WD 5 7/12	0 days
Place & Finish Slab	WD 6 7/15	WD 6 7/15	WD 6 7/15	WD 6 7/15	0 days
Erect Wood Framing	WD 7 7/16	WD 11 7/22	WD 7 7/16	WD 11 7/22	0 days
Complete Rough Framing	WD 12 7/23	WD 13 7/24	WD 13 7/24	WD 14 7/25	1 day
Install Shingles	WD 14 7/25	WD 16 7/29	WD 15 7/26	WD 17 7/30	1 day
Install Doors & Windows	WD 12 7/23	WD 12 7/23	WD 12 7/23	WD 12 7/23	0 days
Install Siding & Cornice	WD 13 7/24	WD 14 7/25	WD 13 7/24	WD 14 7/25	0 days
Rough-in & Trim Electrical	WD 15 7/26	WD 16 7/29	WD 16 7/29	WD 17 7/30	1 day
Paint Exterior	WD 15 7/26	WD 17 7/30	WD 15 7/26	WD 17 7/30	0 days
Final Inpection	WD 18 7/31	WD 18 7/31	WD 18 7/31	WD 18 7/31	0 days

Note: WD= workdays

Advantages and Disadvantages of AOA Diagrams. AOA diagrams use simple end-to-start logic to relate one activity to another, and clearly illustrate which activities must be completed on time to avoid project delays. All activities are shown as beginning at one circle, or node, and ending at another node.

As AOA diagrams are updated, the results affect all critical activities in the diagram. Since all activities are linked together, a change in the duration of one critical activity will change the start and finish dates of the subsequent critical activities when the network calculation is redone.

AOA diagrams can be converted to simplified bar charts to enhance communication with owners and subcontractors. Each activity is placed on a bar chart in the location determined by the network calculations. This allows the advantages of the bar charting method to be combined with the advantages of logic diagrams. You can simplify the schedule information that you share with owners or subs, for example, by not including specific information about float.

One disadvantage of AOA diagrams is that they can become complex and managing them can quite time-consuming. AOA diagrams are harder to draft than bar charts and require a more complete analysis of the project. Because the AOA diagram represents relationships among activities in greater detail, if the diagram is to be useful, each relationship must be correctly drawn. Also, AOA diagram logic is difficult to change, particularly when two activities share a common node.

Making the forward and backward pass manually can be cumbersome. The diagrams can become cluttered with all the numbers displaying the early and late start dates. The calculations are a simple process, but if changes are made to the diagram, the calculations have to be redone from the beginning.

Activity-on-Node (AON) Diagrams

An Activity-on-Node (AON) diagram accomplishes the same things as an Activity-on-Arrow (AOA) diagram, but the information is presented differently. Whereas the AOA diagram uses arrows to represent activities, the AON diagram uses arrows to show *relationships* between activities. An AOA diagram uses nodes to represent event times (points represent the start or stop of activities), but an AON diagram uses the nodes to represent the activities themselves. Typically, the activity description is written in the center of the node with the activity duration noted just below. The sample garage would appear in an AON diagram as shown in Figure 2.14. For ease of comparison between AOA and AON diagrams, the same numbers (5, 10, 15, and so forth) have been assigned to the activities. Remember, the activity numbers, like node numbers, are simply place markers—they bear no relation to the duration calculations.

Figure 2.14: Activity-on-Node Diagram for Garage Schedule

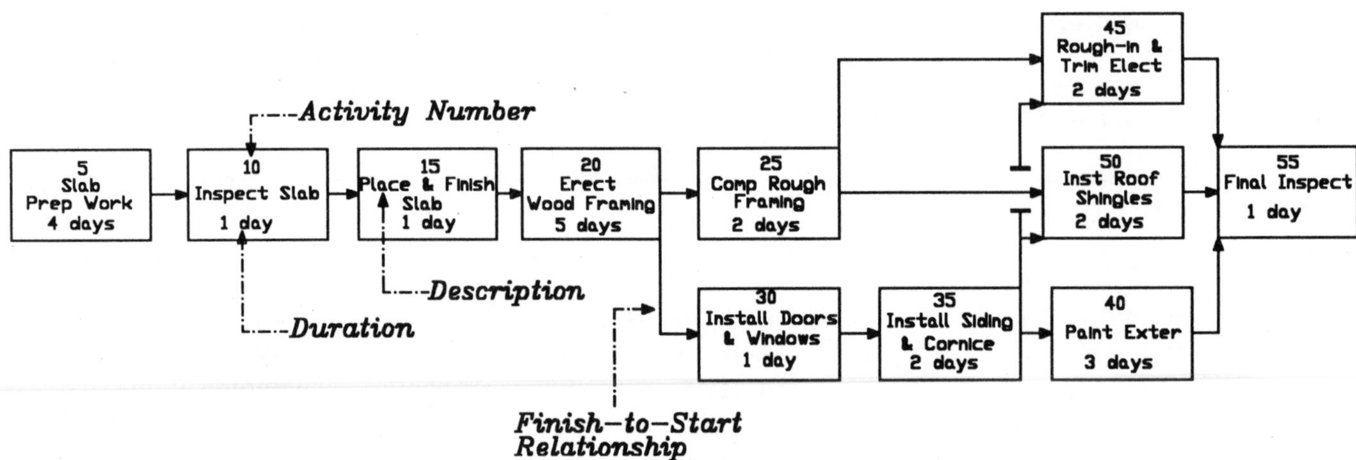

The logic used is the same end-to-start logic used in the AOA diagrams. However, this method allows you to visually display other relationships as well. AON diagrams can be created to show start-to-start and end-to-end relationships. Figure 2.15 identifies the three relationships.

Another feature of the AON diagram is that it allows you to include a preplanned *lag time* or delay in the start of the next activity. The lag time is assigned to the relationship (and so appears below the arrow) between activities. The lag can act as a restraint to the

Figure 2.15: AON Diagram Showing Relationships

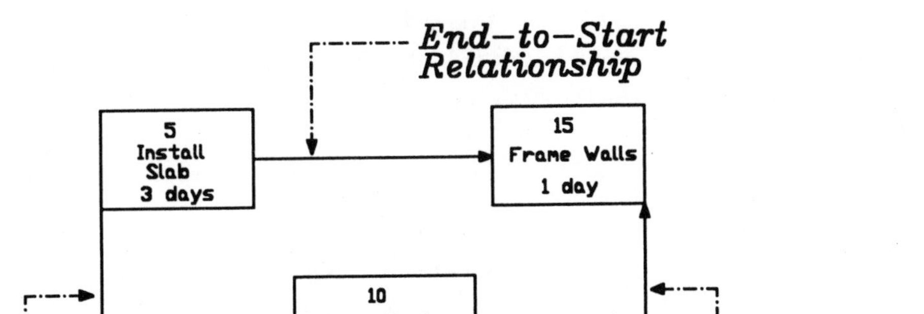

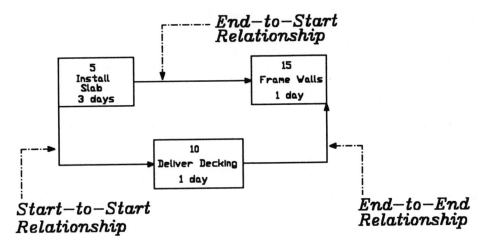

start of the next activity. For example, a slab should cure for seven days before framing is started. No one is responsible for the curing and once the slab is placed, curing will automatically begin, so this need not be recorded as an activity with its own node. However, the seven days must be accounted for in the schedule somehow. This can be handled easily, as shown in Figure 2.16.

Figure 2.16: AON Diagram Showing Lag Time

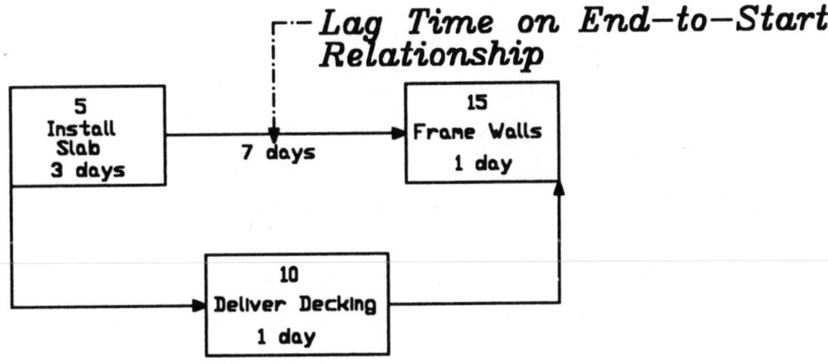

You can use the relationships and lag times to construct a logic diagram for the project. The AON diagram allows you to represent relationships other than the basic end-to-start relationships. Also, AON diagrams can provide for delays in the start of activities and do not need dummy activities.

AON Network Calculations. Network calculations for an AON diagram are similar to the calculations for the AOA diagram. You start at the first activity, add the duration to zero, and obtain the early finish workday. Each early finish workday number then becomes the early start workday for the next activity, until the last activity's start and finish workday numbers have been calculated. When the relationship between two activities includes a lag time, the lag days are added to the early finish workday to establish the early start of the next activity.

Again, the backward pass reverses this process. Starting at the last activity and proceeding backward, you calculate the late finish and late start workday numbers of each activity. When the relationship between two activities contains a lag time, the lag time is subtracted from the late start workday number on the backward pass.

Figure 2.17 shows how forward and backward pass calculations would appear on an AON diagram.

Time-Scaled Logic Diagrams. Another way to display the information presented in a logic diagram is to use a time-scale. Both AOA and AON diagramming methods can be time-scaled. Figure 2.18 shows the activity-on-arrow diagram and Figure 2.19 shows the activity-on-node diagram.

Note that Figure 2.19 is similar to the bar chart in Figure 2.2. The only difference is the display of the relationships. A time-scaled AON diagram allows the presentation of the relationships in a visual format, which gives you instant feedback on the effect of decisions.

Advantages and Disadvantages of AON Diagrams. AON diagrams take the bar chart a step forward by adding logic to a familiar scheduling tool.

AON diagramming is the basis for most inexpensive scheduling software. The software allows the user to place, move, modify, and rearrange the activities without having to manually recalculate and redraw each version. Because such software lends itself to manipulation and "what if" scenarios, it can be useful during the initial planning of a project.

Figure 2.20 shows the garage schedule as generated using scheduling software. The schedule shows the activities in a time-scale. Although nonworkdays—weekends and holidays—are still shown, activities that span a weekend are extended for the proper number of workdays. If you use a computer, you can create the diagram, complete the network calculation, and generate printouts in a matter of minutes. (Note the "date printed" automatically included by the software in Figure 2.20.)

Like AOA diagrams, as AON diagrams are updated, the results will automatically affect all critical activities in the diagram. Updating the schedule as the project progresses may cause activities to change their start dates, depending on the length of delays and whether activities are critical or not.

AON diagrams can be converted to bar charts to communicate information to all parties. This allows the advantages of the bar charting method to be included with logic diagrams.

Like AOA diagrams, however, AON diagrams can be complex and time-consuming to generate. This is especially true if you are scheduling by hand.

AON diagrams are harder to draft than other types of diagrams and—because they incorporate lag times—require a more complete analysis of the project.

Like AOA diagrams, AON diagrams require forward and backward pass calculations, which can be cumbersome.

Figure 2.17: AON Diagram Showing Network Calculations

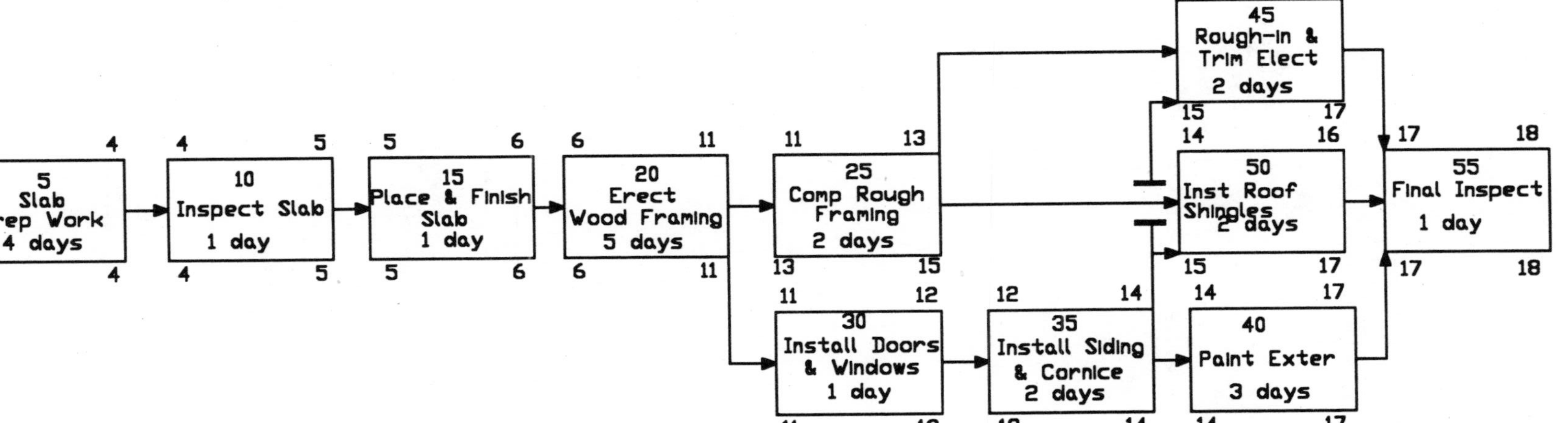

Activity Description	EST	+	Duration	=	EFT	LFT	-	Duration	=	LST
Slab Prep Work	0	+	4	=	4	4	-	4	=	0
Inspect Slab	4	+	1	=	5	5	-	1	=	4
Place & Finish Slab	5	+	1	=	6	6	-	1	=	5
Erect Wood Framing	6	+	5	=	11	11	-	5	=	6
Comp Rough Framing	11	+	2	=	13	15	-	2	=	13
Inst Doors & Windows	11	+	1	=	12	12	-	1	=	11
Inst Siding & Cornice	12	+	2	=	14	14	-	2	=	12
Install Roof Shingles	14	+	2	=	16	17	-	2	=	15
Paint Exterior	14	+	3	=	17	17	-	3	=	14
Rough-in & Trim Elec	14	+	2	=	16	17	-	2	=	15
Final Inspection	17	+	1	=	18	18	-	1	=	17

Figure 2.18: Time-Scaled AOA Diagram for Garage

Work Days	1	2	3	4	5	6	7	8	9	10	11	12	13	14	15	16	17	18
Cal. Days	7/8	7/9	7/10	7/11	7/12	7/15	7/16	7/17	7/18	7/19	7/22	7/23	7/24	7/25	7/26	7/29	7/30	7/31

Slab Prep Work
4 days
Inspect Slab
1 day
Place & Finish Slab
1 day
Erect Wood Framing
5 days
Comp Rough Framing
2 days
Rough-In & Trim Electrical
2 days
Inst Roof Shingles
2 days
Final Inspect
1 day
Install Doors & Windows
1 day
Install Siding & Cornice
2 days
Paint Exterior
3 days
5 10 15 20 25 30 35 40 45 50 55

Legend
Activity Line
Dummy Line
Float Line

Figure 2.19: Time-Scaled AON Diagram for Garage

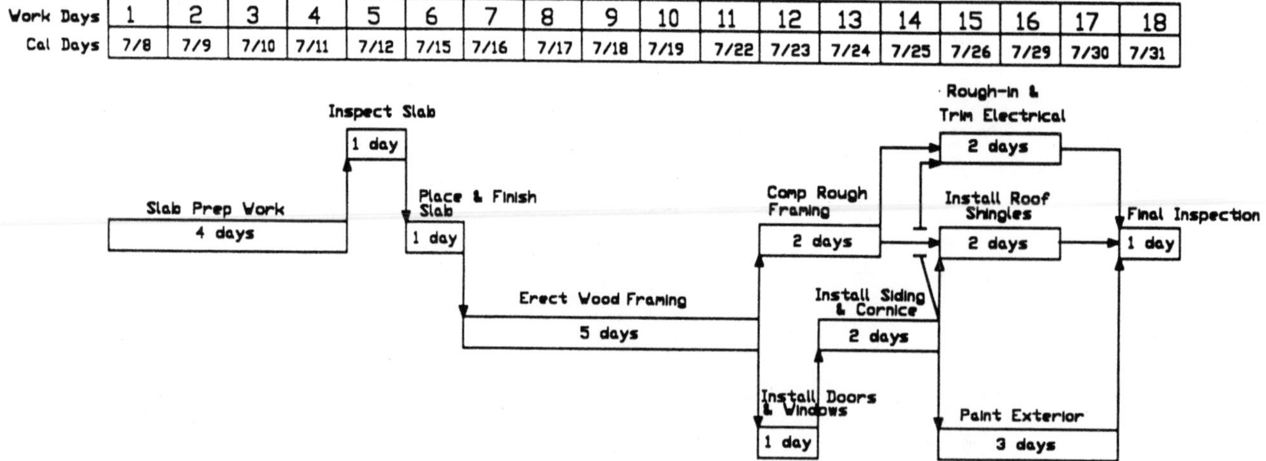

Figure 2.20: Computer-Generated Schedule for Garage

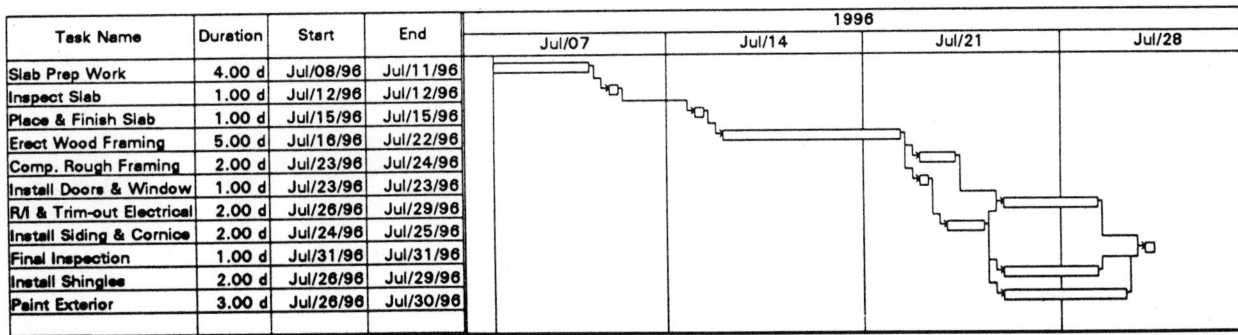

Task Name	Duration	Start	End
Slab Prep Work	4.00 d	Jul/08/96	Jul/11/96
Inspect Slab	1.00 d	Jul/12/96	Jul/12/96
Place & Finish Slab	1.00 d	Jul/15/96	Jul/15/96
Erect Wood Framing	5.00 d	Jul/16/96	Jul/22/96
Comp. Rough Framing	2.00 d	Jul/23/96	Jul/24/96
Install Doors & Window	1.00 d	Jul/23/96	Jul/23/96
R/I & Trim-out Electrical	2.00 d	Jul/26/96	Jul/29/96
Install Siding & Cornice	2.00 d	Jul/24/96	Jul/25/96
Final Inspection	1.00 d	Jul/31/96	Jul/31/96
Install Shingles	2.00 d	Jul/26/96	Jul/29/96
Paint Exterior	3.00 d	Jul/26/96	Jul/30/96

Printed: Mar/06/95
Page 1

SUMMARY POINTS AND TIPS

- Bar charts and logic diagrams present similar information, and logic diagrams usually can be used to create bar charts with little effort. However, bar charts do not contain all of the information needed to adequately manage a complex project.
- It is not the specific scheduling method that is important, but the combination of accurate planning and thorough communication of what is expected and needed to complete the project. As builders and remodelers depend more on subcontractors, better communication is required to ensure that all parties can plan how to assign their resources.
- Don't develop your project schedule in the dark. Ask the subcontractors and material suppliers the right questions: When? How long? Does your work affect anyone else?

Chapter 3

Scheduling a Residential Project

So far, we have discussed the following steps in developing a schedule: establishing a plan; developing a diagram; performing network analysis; assigning information to activities; and creating graphics and reports to enhance communications with workers, subcontractors, customers, and others. This chapter will apply these steps to the development of a schedule for a sample house. A floor plan and front elevation for the sample house appear in Appendix A and a summary estimate is located in Appendix B. The plan, elevation, and estimate each provide information that helps determine the activities, resources, and costs that will go into the schedule.

The sample house has 3,350 square feet of living space and is constructed on a slab-on-grade foundation with wood stud framing, conventional rafter roof system, and drywall interior with paint and vinyl wallpaper. The exterior has brick, synthetic stucco, and wood siding. The main house has four bedrooms with a carport. Attached to the carport is a storage area.

The ceiling heights vary from nine feet to ten feet. Three ceiling designs are used in this house—vaulted, cathedral, and layered designs—which will complicate the roof framing. The roof is hipped and gabled with a dutch hip at one end. The fascia and soffit detail shows a sloping fascia board with continuous soffit vents.

"We" will be the builder for this example. We will subcontract the sitework for this project. The sitework includes driveways, sidewalks, and landscaping. The estimate includes this information and reflects the total work required by the contract.

The detailed version of the project estimate would provide information concerning the finish items required to build the house. There are several different types of floor coverings. Imported marble is to be used in the entry area and clay tile will be used in the kitchen, breakfast area, living room, and dining room. Carpet will be used in the bedrooms and hall areas and ceramic tile in the bathrooms.

Interior trim materials and doors and windows are identified. Note that some windows contain arched tops, which are special-order items. The use of eleven window sizes will require extra attention to ensure proper fit and placement.

For the purposes of this example, it is assumed that the estimate is complete except for wallpaper and color selections that the owner must make, and that the contract is based only on the items identified in the estimate. To keep things simple, we will also assume that owner-selected items will not change the contract.

THE PLAN

The first scheduling step is to plan the project. During this stage, we will select the activities (using the master list established in Chapter 1), establish relationships with other activities, and organize the work.

To keep focused on the activities that are important, use your detailed estimate for the project to select activities from the master list. Depending on the scope and complexity of the project, your final project list may include a few additional activities; alternatively, it may use only a portion of the activities on the master list. Proceed down the project list and determine the duration and cost of each activity according to your detailed estimate. (The estimates for the house and remodel projects presented in Appendix B and D are summaries. In creating the detailed plans for the project phases that appear in the chapters the author has broken out the costs and durations that would appear in a more detailed version of the estimates.)

Remember: this is a flexible listing. Activities are determined by a builder's or remodeler's priorities and decisions concerning what is to be done and who is to do it. Each builder or remodeler will create and recreate this list a little differently, breaking down some activities into more detail or deleting others, during the process of creating the final project list.

Site and Slab Phase

The first two phases of the breakdown list will be the project startup and sitework. These activities are standard for most of our projects. Our site subcontractor (Site Sub) has been used on several projects, has always performed above-average work, and has always been on time.

Each activity in the list is followed by notations about its duration, estimated cost, and who is responsible. The activities for these two phases appear in Figure 3.1.

Figure 3.1: Plan for Startup and Sitework

Project Startup	
Obtain Permit	3 days, $500, Builder
Obtain Temporary Power	2 days, $50, Electrician
Obtain Temporary Water	2 days, $25, Plumber
Obtain Builders Risk Insurance	1 day, $455, Builder
Surveyor to Locate Building Corners	3 days, $250, Surveyor
Initial Sitework	
Site Grading	5 days, $2,700, Site Sub
Finish Sitework	
Grade Driveway & Walks	½ day, $250, Site Sub
Install Concrete Driveway	4 days, $5,000, Site Sub
Form and Place Sidewalks	1 day, $1,056, Site Sub
Site Landscaping	3 days, $2,500, Landscaper

The assigned durations for these activities are based on our company's historical experience and on discussions with the subcontractors involved with this project. Hint: ask all your subcontractors to provide a time estimate for their work. If you must create the schedule before selecting all your subcontractors, you will have to establish the durations.

Some builders may believe the durations we have assigned to specific activities are too long or too short. The important thing to remember is that the durations must be realistic and also must reflect your company's priorities in terms of time, money, and allocation of personnel. Site grading requires time to move equipment to and from the site, time to complete the work, and additional time for possible delays. Durations for activities that can be delayed by the weather should include some extra time to allow for that possibility. Our five-day duration for site grading includes two weather days. Duration calculations will be discussed in more depth later in this chapter.

This house has a slab-on-grade foundation. Therefore, activities such as those having to do with foundation block that appear on our company's master list under Foundation Work will not be used for this schedule. We typically install slabs using our own crew.

The complicated layout of this house will lead us to schedule installation of the slab in three separate placements—the main house, the garage/storage area, and the concrete patio. A detailed analysis is only needed for the house slab; the other two placements may be summarized.

The activities for the slab-on-grade phase are shown in Figure 3.2.

Figure 3.2: Plan for Slab Installation

Activity	Duration, Cost, Responsibility
Slab-on-Grade Work	
Utility Rough-in	
Plumbing Rough-in	3 days, $2,175,Plumber
HVAC Rough-in	½ day, $275, AC Sub
Radon System Rough-in	½ day, $250, Radon Sub
Slab Preparation Work	
Lay Out and Set Batter Boards	1 day, $300, Builder
Set Edge Forms	1½ days, $1,026, Builder
Excavate Turndown Edge	½ day, $285, Builder
Spread Gravel	½ day, $552, Builder
Termite Treatment	½ day, $1,000, Termite Sub
Place Poly and Welded Wire Mesh	½ day, $1,056, Builder
Set Grade Stakes	½ day, $50, Builder
Place Slab	
Inspect Slab	½ day, $0, Inspector
Place & Finish Concrete	1 day, $3,952, Builder
Place Garage/Storage Slab	3 days, $1,934, Builder
Place Concrete Patio	2 days, $925, Builder

The activities we list for installing the garage/storage area and the concrete patio take into consideration all the individual activities we have defined for the main house. (The builder determines the level of detail. Summarizing helps reduce the number of activities to keep lists manageable.)

In Figure 3.2, many activities are located under the category Slab Preparation Work. To simplify the list in this category, we can combine the "builder" activities into a single activity called Builder Slab Preparation, with a duration of four and one half days and a cost of $3,269. Further combination of certain activities—Lay Out and Set Batter Boards, Set Edge Forms, Excavate Turndown Edge—can save time, possibly reducing the duration of these activities.

Note that when combining our activities, the work of the termite subcontractor has been kept separate. Only one person should be listed as responsible for each activity on the schedule.

The summary activities contain the material and labor costs required to complete the installation. The cost is determined by adding up all the items associated with this work. For example, the $925 cost for placing the concrete patio was obtained by estimating a portion of the total costs for concrete, reinforcement, other materials, and labor that appear in the project estimate in Appendix B.

Under Utility Rough-in, the air conditioning subcontractor (AC Sub) will have a run of conduit for the installation of the refrigerant lines to the inside unit and a run of duct for the downdraft range in the kitchen. Each of these activities may require less than a half day of work, but a half day should be the smallest time frame assigned to an activity.

The estimate becomes the basis of the schedule and plan. Any estimating method that you use needs to provide information about how you plan to build the project. Square foot cost, detailed unit cost, or detailed productivity-based estimating are all accepted estimating methods. However, only the productivity-based method will help determine accurate durations. This estimating method also takes the most time and requires the most detailed tracking of the actual work.

A detailed item list helps builders project durations and costs for specific activities based on productivity of the assigned crew. One example could be installing vapor barrier. The crew selected consists of one carpenter who is paid at the rate of $18.75 per hour. This rate includes wages, insurance, and taxes. Historically, the carpenter has been able to place 1,000 square feet of vapor barrier per hour. To install the vapor barrier in 3,525 square feet will require 3.52 hours, rounded up to one half day (four hours). The labor cost for this item of work would be 4 X $18.75, or $75.00.

The key to using the productivity estimate is maintaining accurate productivity factors (such as the 1,000 square feet per hour). The builder can determine the factors from company records or use published information. These productivity factors should be used as a guide for determining a *reasonable* duration for an activity; they are only a starting point. The builder should also involve the subcontractor or superintendent before deciding on the final duration.

Builders who estimate by unit price have no initial guide as to the time required to complete an activity. If the vapor barrier example used above costs $0.02 per square foot to install, the labor cost (rounded) would come to $71.00. The builder still must determine a duration by guessing, by determining a productivity factor, or by relying on the subcontractor's time estimate.

Many builders mentally calculate durations based on their knowledge of their crews and experience with similar projects.

Site and Slab Phase Diagram

It is a good idea to develop the overall diagram in sections, particularly if you have many activities to consider. Once the site and slab activities have been identified, a diagram can be created for these phases.

Figure 3.3 presents one possible way to diagram a schedule for this work. The sequence followed in the logic diagram is based on the following considerations and restraints. The words that appear in boldface indicate a reference to an activity or activities listed in Figure 3.1 or Figure 3.2.

1. The builder **obtains the building permit.**
2. The site subcontractor arrives at the site and performs the **site grading.**
3. The plumber and electrician begin to set up the **temporary utilities** at the same time the site sub is working.
4. The surveyor then **locates the building corners** for the layout to begin. The surveyor must wait until the site is graded.
5. Slab preparation work does not begin until temporary power is provided and the surveyor is finished.
6. **Excavate turndown, plumbing rough-in, HVAC rough-in,** and **radon system rough-in** can begin after the **lay out and batter boards** and the **edge forms** activities are completed.
7. We have required the plumber to finish the temporary water before beginning the plumbing rough-in. This ensures a water source for testing and maintaining pressure on the water pipes during the pour.
8. Once the underground rough-in work is in place, the crew can **spread gravel, treat for termites,** and then **install poly, welded wire mesh, rebar,** and **grade stakes.**
9. After the slab prep work is complete, the **inspection** can take place. This time should also be used to check the framework layout for accurate dimensions and square corners. (You will need to check the inspection requirements in your area; these vary from place to place.)
10. **Placing and finishing** the concrete slab is the last activity in this phase for the main house.
11. The **garage/storage area** and **patio** slabs can be installed after the main house slab is finished.
12. The **Finish Sitework** activities will occur later in the project.

Rough Framing and Dry-in

After the slab is completed, the structure phase can begin with rough framing. The framing will include all walls, ceiling joists, rafters, and decking required to complete the structure of the house. Once the rough framing is complete, dry-in activities can begin.

Note: We have developed this example using stick framing techniques, but some builders might elect to use manufactured components. Each technique has advantages and disadvantages. Manufactured components require more lead time and shop drawings, but can reduce jobsite erection cost and time. The stick framing can begin with an overnight delivery of framing material. A good, experienced framing crew can stick build the most complicated house quickly and with a minimum of waste. On the other hand, the manufactured components can be erected easily with good supervision and a less specialized crew. You can select whichever method best fits your goals.

Figure 3.3: Partial Site and Slab Phase Diagram

Phase/Activity List	Resp	Duration	Start	End
Project Startup Phase		11.00 d	Sep/09/96	Sep/23/96
Obtain Permit	Builder	3.00 d	Sep/09/96	Sep/11/96
Obtain Temporary Power	Electrician	2.00 d	Sep/12/96	Sep/13/96
Obtain Temporary Water	Plumber	2.00 d	Sep/12/96	Sep/13/96
Surveyor Locate Bldg Corners	Surveyor	3.00 d	Sep/19/96	Sep/23/96
Sitework Phase		5.00 d	Sep/12/96	Sep/18/96
Site Grading	Site Sub	5.00 d	Sep/12/96	Sep/18/96
Slab-on-Grade Phase		12.00 d	Sep/24/96	Oct/09/96
Lay Out and Set Batterboards	Builder	1.00 d	Sep/24/96	Sep/24/96
Excavate Turndown Edge	Builder	0.50 d	Sep/25/96	Sep/25/96
Set Edge Forms	Builder	1.50 d	Sep/25/96	Sep/26/96
HVAC Rough-in	AC Sub	0.50 d	Sep/25/96	Sep/25/96
Radon System Rough-in	Radon Sub	0.50 d	Sep/25/96	Sep/25/96
Spread Gravel	Builder	0.50 d	Sep/30/96	Sep/30/96
Plumbing Rough-in	Plumber	3.00 d	Sep/25/96	Sep/27/96
Termite Treatment	Termite	0.50 d	Sep/30/96	Sep/30/96
Set Grade Stakes	Builder	0.50 d	Oct/01/96	Oct/01/96
Place Poly and WWM	Builder	0.50 d	Oct/01/96	Oct/01/96
Inspect Slab	Inspector	0.50 d	Oct/01/96	Oct/01/96
Place & Finish Concrete	Builder	1.00 d	Oct/02/96	Oct/02/96
Place Garage/Storage Slab	Builder	3.00 d	Oct/03/96	Oct/07/96
Place Concrete Patio	Builder	2.00 d	Oct/08/96	Oct/09/96

1996

Sep/08 | Sep/15 | Sep/22 | Sep/29 | Oct/0

Shaded activities are on the critical path.

Dry-in should begin as soon as possible after the framing is complete. Dry-in consists of the exterior sheathing, roofing felt, and flashing, and exterior doors and windows. These items help protect the framing from water damage and excess moisture absorption.

For our sample house, we will again use a framing subcontractor to perform the rough framing activities. If we were to use our own crew for these items of work, the specific breakdown of activities would vary but the basic sequence and concepts would be the same.

The framing sub has one lead carpenter, a helper, and four framers, who can perform the rough and finish framing work such as installing siding, exterior doors, windows, fascia, and soffit. The synthetic stucco and roofing subcontractors have worked with us on several projects. They perform to our standards and understand our system, and should be an asset to the timely completion of this project.

Although our mason has been recommended by other contractors, this sub has not worked for us in the past. In this situation we would make a mental note that some additional breakdown activities may be required to properly schedule this sub's work.

The exterior of our project involves three different finish systems. Brick covers most of the exterior wall area. Synthetic stucco is used to enhance the appearance of the front entryway and to provide a contrast to the brick. Wood siding is used in covered areas and in the rear locations.

The next phase of our schedule includes drywall and insulation materials that are easily damaged by moisture. This work can be started before the exterior finish items are completed. However, you then run the risk of moisture damage to the insulation and drywall. This decision should therefore be made onsite at the proper time and should not be fixed in the schedule.

The activities in this phase require several inspections, both by the builder and by local authorities.

The framing inspection should verify the wall locations and rough opening sizes. The project should also be checked for square and plumb walls. The electrical rough-in should be checked to make sure all switch boxes, outlets, and fixtures are in the proper locations. Plumbing should be checked for water pipes running to all locations such as toilets, sinks, refrigerator line, and other areas.

The activities required for the structure and dry-in phases are outlined in Figure 3.4.

The framing subcontractor has the most activities in this section and the durations of the activities may need to be adjusted for weather. Rain or adverse weather would delay these activities and possibly the rest of the project.

The delivery of the framing and miscellaneous material has been assigned a one-day duration. This time frame assumes all material is available locally and delivery can be easily accomplished. To reduce loss or damage of materials, most builders and remodelers plan to have the materials delivered as close to the actual installation date as possible. It is also critical to be sure the materials arrive on time: having a crew wait on materials is costly.

In the activity list above, the cost assigned to delivery activities is the material cost, not the labor. The wall and roof framing activities contain amounts for the labor cost. The labor cost is divided so that the wall framing is assigned one third of the cost while the roof and ceiling framing are assigned two thirds. Developing a standard ratio or value for items like framing is a good practice.

Our delivery of the specialty tubs must be timed so that they are in place before the walls go up. Generally, the builder and owner will have selected such fixtures before the

Figure 3.4: Plan for Structure and Dry-in

Task	Duration, Cost, Responsibility
Structure Phase	
Erect Rough Framing	
Order Framing Material	1 day, $0, Builder
Deliver Framing Material	2 days, $0, Builder
Walls and Sheathing	1½ days, $7,111, Framing Sub
Roof Framing and Decking	2½ days, $4,197, Framing Sub
Complete Rough Framing	
Check Plumb and Square	½ day, $150, Builder
Check Door Openings (size & square) with Above Item	
Install Deadwood and Blocking	½ day, $150, Framing Sub
Wall Insulation	1 day, $836, Insulator
Framing Inspection	½ day, $0, Inspector
Clean Up Waste Material	½ day, $250, Builder
Utility Rough-in	
Order & Deliver Tubs	4 days, $2,500, Builder
Plumbing Top-out	4 days, $2,013, Plumber
Plumbing Inspection	½ day, $0, Inspector
HVAC Rough-in	3 days, $3,347, AC Sub
Electrical Rough-in	2 days, $4,184, Electrician
Electrical Inspection	½ day, $0, Inspector
Structure Dry-in Phase	
Exterior Wood Items	
Install Fascia and Soffits	4 days, $1,871, Framing Sub
Order Exterior Doors & Windows	½ day, $0, Builder
Deliver Exterior Doors & Windows	10 days, $0, Builder
Install Exterior Doors & Windows	4 days, $435, Framing Sub
Install Overhead Door	½ day, $410, Framing Sub
Install Siding	½ day, $368, Framing Sub
Exterior Masonry Items	
Order Brick	1 day, $0, Builder
Deliver Brick	15 days, $3,606, Builder
Install Brick	4 days, $4,287, Mason
Roofing Items	
Install Shingles	3 days, $4,037, Roofer
Exterior Synthetic Stucco System	
Set Panels and Trim	½ day, $1,000, Stucco Sub
Apply Finish	3 days, $1,371, Stucco Sub
Special Custom Items	
Install Fireplace	1 day, $3,076, Fire Sub
Install Chimney Cap	½ day, $500, Fire Sub

slab is placed. The plumber has to run water and waste lines to the fixtures.

Some builders ask the home buyer or owner to assume certain responsibilities, which should be translated to activities on the schedule. Owner activities can include the selection of materials, paint colors, and special fixtures. The owner must be given deadlines for making the selections. Failure to communicate the importance of such deadlines to the owner can cause delays and/or increase project costs.

In our scenario, the AC subcontractor will receive one third of the cost of the HVAC contract for installing the ducts and rough-in for the equipment. The plan allows three days for the crew to bring materials to the project, install materials, clean up, and remove trash. The sub should understand not only when the activity should start, but also the scope of what is expected within the time allowed.

For a schedule to be realistic, the subcontractors must have some say regarding the amount of time allotted to their activities. Remember, subcontractors have other projects and must keep their employees working at all times to make a living. A task with a three-day duration may require two employees. Shortening the duration could require more employees who may not be available when needed.

For our sample house we have assumed that the mason is the only subcontractor we have not used on a previous project. As the builder, we will be providing all materials and the subcontractor will provide the labor to do the brick work. For better tracking, we have split the masonry into three activities—order, deliver, and install brick. The brick order time allows the selection to be made far enough in advance to avoid any delay. For clarity, material costs are assigned to delivery and labor costs are assigned to installation.

The roofer has agreed to complete the shingles, felt, and flashing in three days. The roofer originally estimated that the fifty four squares of shingles will take 86.4 labor hours or eleven workdays to complete using only one roofer. If the roofer can put four people on the project, however, the time can be compressed to 2.7 days. The roofer has agreed to this, so we round up the duration to three days.

The fireplace has also been included in the grouping of activities. The framing and stucco cannot be completed without the fireplace. Remember, special items like this need to be installed in a timely manner. Try to save time up front in the planning stage by not leaving out items in the schedule. When activities are forgotten or performed out of sequence, crews may have to return later to redo work or finish such "forgotten tasks." If this happens, it generally costs more in time and money than the initial investment in planning would have cost.

Structure and Dry-in Diagram

Let's construct the diagram for the structure and dry-in phases. One possible diagram is shown in Figure 3.5. Because this is a partial diagram, preceding activities are not shown. The words appearing in boldface indicate a reference to an activity or activities listed in Figure 3.4. This logic diagram is based on the following considerations and restraints:

1. The initial activities are the **order and delivery of the framing, brick, and tubs**. The placing of the order for materials can be done as early as possible. Therefore, delivery activities should not become critical.
2. The **walls and sheathing** and the **fireplace installation** can begin after the material arrives. The **roof framing and deck** will follow the walls. The fireplace should be completed once the framing is finished.

3. The **fascia and soffit**, the **deadwood and blocking**, and the checking will begin as soon as the roof decking is in place. **Checking and inspections** should be made as quickly as possible. Corrections should be taken care of promptly.
4. The **plumbing top-out, HVAC rough-in**, and **electrical rough-in** can begin after the framing is checked. These subcontractors need to be sequenced to avoid having too many people trying to work in the same place. The plumber has more stringent codes and usually starts first. The HVAC and electrical subcontractors will follow the plumbing.
5. After the framing crew finishes the fascia and soffit, they can then install the **exterior doors, windows, overhead door,** and the **exterior siding**.
6. The roofer begins as soon as possible to avoid damage to the decking and framing. Framing crews sometimes cover the deck with felt. We have included this work in the roofer's contract and here it is included in the shingle activity, although it is often part of the framer's contract. **Shingles** will be installed after the installation of the vent stacks, fascia, soffit, felt, and flashing are completed.
7. The building will be weather-tight after the roofing shingles, doors, and windows are installed. The **wall insulation** and **cleanup** will start after the roofing and electrical inspections. The **city inspection** for framing takes place after the builder is satisfied that all work meets specifications. Some inspectors want to see the job before you install wall insulation. Be sure to check all local requirements before you schedule the job.
8. After the doors, windows, and siding, the **brick** and **synthetic stucco** can be installed. Because the brick and synthetic stucco for this house are not in the same area (see elevation in Appendix A), we will assume that both subcontractors can work on the exterior at the same time. In places where they abut, it doesn't matter who gets there first. Sometimes it is better to schedule subcontractors at different times even though they can work in the same space at the same time. When people work in tight conditions, productivity decreases, damage to the project can occur, and the risk of personal injury increases.
9. Once the synthetic stucco has been finished around the chimney framing, the **chimney cap** can be installed to seal the top of the chase against any water leaking into the building.

Finish Phase

Finishing can proceed when the framing and rough-ins are complete and inspected, the building is dried-in, and insulation is in place. The drywall installation is the first activity in the finish phase of the project. According to our plan, drywall installation begins after the exterior is completed and the city inspection department has approved the building. Scheduling it this way helps us avoid damage to the sheetrock caused by moisture coming through the walls. To further protect the sheetrock, we will advise workers to close the doors and windows at the end of each workday.

The finish phase can be complex, and the planned schedule can create problems at several points. The variety of materials and subcontractors typically involved in the finish phase dictates that we carefully review the activities and their sequence for every project we schedule.

Our plan includes subcontracting the sheetrock, painting, interior trim, and wallpaper. Our interior carpentry subcontractor will install the interior trim, shelving, hardware, and doors.

Figure 3.5: Partial Structure and Dry-in Phase Diagram

Phase/Activity List	Resp	Duration	Start	End
Structure Phase		21.50 d	Sep/23/96	Oct/23/96
Order Framing Material	Builder	1.00 d	Sep/23/96	Sep/23/96
Deliver Framing Material	Builder	2.00 d	Sep/24/96	Sep/25/96
Wall and Sheathing	Framing	1.50 d	Oct/04/96	Oct/07/96
Roof Framing & Decking	Framing	2.50 d	Oct/07/96	Oct/09/96
Select Whirlpool Tub	Owner	1.00 d	Sep/23/96	Sep/23/96
Order & Deliver Tubs	Builder	4.00 d	Sep/24/96	Sep/27/96
Plumbing Top-out	Plumber	4.00 d	Oct/10/96	Oct/16/96
Plumbing Test & Insp	Plumber	0.50 d	Oct/17/96	Oct/17/96
HVAC Rough-in	AC Sub	3.00 d	Oct/10/96	Oct/15/96
Radon System Top-out	Radon Sub	1.00 d	Oct/18/96	Oct/18/96
Inst Fireplace	Fire Sub	1.00 d	Oct/07/96	Oct/08/96
Electrical Rough-in	Electrician	2.00 d	Oct/16/96	Oct/17/96
Electrical Inspection	Electrician	0.50 d	Oct/18/96	Oct/18/96
Inst Deadwood and Blk	Framing	0.50 d	Oct/10/96	Oct/10/96
Check Plumb & Square	Builder	0.50 d	Oct/10/96	Oct/10/96
Framing Inspection	Builder	0.50 d	Oct/10/96	Oct/10/96
Clean up All Material	Builder	0.50 d	Oct/21/96	Oct/21/96
City Inspection	Inspector	0.50 d	Oct/21/96	Oct/21/96
Wall Insulation	Insulator	1.00 d	Oct/22/96	Oct/23/96
Structure Dry-in Phase		22.50 d	Sep/23/96	Oct/24/96
Install Fascia and Soffits	Framing	4.00 d	Oct/10/96	Oct/17/96
Install Shingles	Roofer	3.00 d	Oct/17/96	Oct/22/96
Order Ext Doors & Win	Builder	0.50 d	Sep/23/96	Sep/23/96
Deliver Ext Doors & Win	Builder	10.00 d	Sep/23/96	Oct/07/96
Install Ext Doors & Win	Framing	4.00 d	Oct/10/96	Oct/17/96
Install Overhead Door	Framing	0.50 d	Oct/17/96	Oct/17/96
Install Siding	Framing	0.50 d	Oct/18/96	Oct/18/96
Select Brick	Owner	1.00 d	Sep/23/96	Sep/23/96
Order Brick	Builder	1.00 d	Sep/24/96	Sep/24/96
Deliver Brick	Builder	15.00 d	Sep/25/96	Oct/16/96
Install Brick	Mason	4.00 d	Oct/18/96	Oct/24/96
Deliver Synthetic Stucco	Stucco Sub	0.50 d	Sep/23/96	Sep/23/96
Set Panels and Trim	Stucco Sub	0.50 d	Oct/17/96	Oct/17/96
Apply Finish	Stucco Sub	3.00 d	Oct/18/96	Oct/22/96
Inst Chimney Cap	Fire Sub	0.50 d	Oct/23/96	Oct/23/96

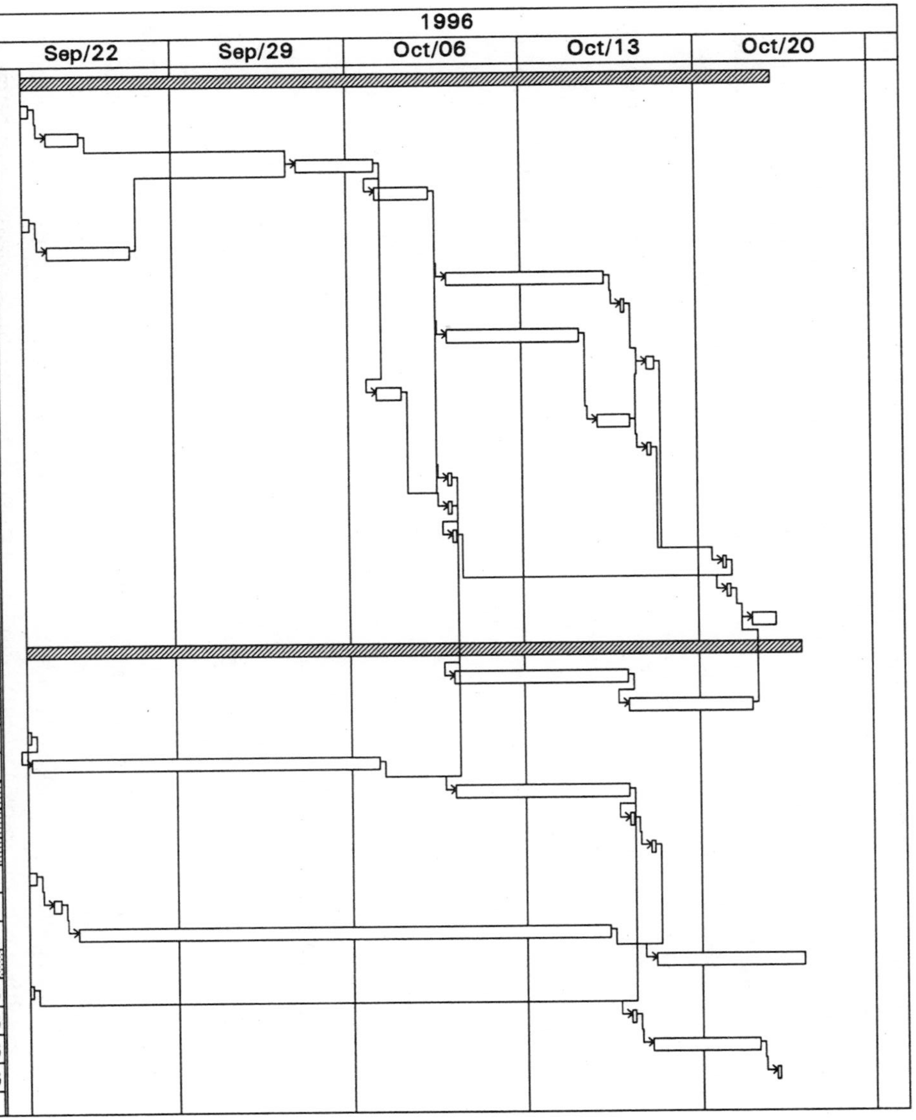

1996

Sep/22 Sep/29 Oct/06 Oct/13 Oct/20

Shaded activities are on the critical path.

The painting follows the trim work, then the floor finishes are installed. The project requires carpet, vinyl tile, clay tile, ceramic tile, and special imported marble. A flooring subcontractor will handle all of this material and installation. The sub determines that the owner-selected marble will take two months to arrive and is nonreturnable.

This schedule must include several miscellaneous finish items such as blown-in attic insulation, measuring for cabinets, bathroom accessories, and setting appliances. The cabinet subcontractor will handle the cabinets and countertops. We will use one of our own carpenters to install the appliances and bathroom accessories.

The utility trim-outs will need to be finished. The plumber will set fixtures, attach water cut-off valves, install faucets, and connect water. The AC subcontractor will set the equipment and install grills and thermostats. The electrician will connect all equipment, complete panels, install light fixtures and devices, and hook up some temporary power to the panel for testing.

The activities listed in Figure 3.6 outline the finish phase.

Finish Phase Considerations

Before we discuss the logic needed to create the logic diagram for this section, there are several considerations to be discussed concerning the finish operation.

The drywall subcontractor has three major activities to perform: hanging the sheetrock, taping and finishing, and cleanup. Generally, taping and finishing will take a minimum of three days. The subcontractor's crew may be able to work in half-day increments to hang the sheetrock, but each section of wall has to dry before the next operation can begin. Therefore, taping and finishing of the sheetrock will take three or more days to complete whether the job is large or small.

Because we require all subcontractors to clean up their own waste material, our plan allows time for this to be accomplished. Cleanup of the sheetrock means both removing all leftover material and scraping the dropped sheetrock mud off the floor. Subcontractors who are required to clean up after themselves most often perform with a minimum of waste.

After the drywall is finished, painting and installation of interior trim and doors takes place. Coordination of these activities is crucial. The painter may wish to paint the trim before it is on the wall. If the trim is stained, the walls should be painted before the trim is installed. If the trim is painted, the painter may want the trim installed first.

During this phase, our trim subcontractor will install the interior trim and doors and perform miscellaneous activities such as installing the shelving and door hardware. The installation of the appliances will be done by our carpenter.

The finish flooring should be done as late as possible to avoid damage to the surfaces. However, the bathroom ceramic tile has to be set before the toilets can be installed. The installation of cabinets and finishing of the floors also need to be coordinated. Which comes first depends on the project's specific circumstances and materials, and on the preference of the builder.

Remember that imported marble? Because of the delivery lag time of forty days, we have to make sure it is ordered early enough to avoid delaying completion of the project. Setting up activities for deliveries helps you to identify items with long delivery or order times early in the project and prevent miscommunication, errors, or damage that can cause long delays. For the partial schedule we are reducing the delivery time to twenty-five days to avoid causing a long delay at the end of the diagram.

Figure 3.6: Plan for Finish Phase

Task	Duration, Cost, Responsibility
Finish Phase	
Interior Wall Material	
Order Sheetrock	1 day, $0, Builder
Deliver Sheetrock	1 day, $2,466, Builder
Hang Sheetrock	2 days, $500, Drywaller
Tape and Finish Sheetrock	4 days, $1,444, Drywaller
Clean Up Sheetrock Waste	1 day, $200, Drywaller
Wall Finish Items	
Paint Interior	4 days, $1,970, Painter
Paint Exterior	4 days, $1,550, Painter
Order Wallpaper	1 day, $0, Builder
Deliver Wallpaper	9 days, $1,000, Builder
Hang Wallpaper	4 days, $1,140, Painter
Interior Wood Trim	
Order & Deliver Interior Trim	2 days, $500, Builder
Install Interior Trim	2 days, $400, Int Trim Sub
Install Closet Shelving & Rods	2 days, $389, Int Trim Sub
Install Interior Doors & Hardware	2 days, $3,386, Int Trim Sub
Finish Flooring	
Install Ceramic Tile	2 days, $1,135, Floor Sub
Install Vinyl Tile	1 day, $204, Floor Sub
Install Clay Tile	5 days, $4,448, Floor Sub
Order Marble	1 day, $0, Floor Sub
Deliver Marble	40 days, $1,000, Floor Sub
Install Marble	2 days, $200, Floor Sub
Install Carpet	2 days, $2,001, Floor Sub
Install Miscellaneous Items	
Insulate Attic	½ day, $1,200, Insulator
Order Cabinets	1 day, $0, Builder
Build & Deliver Cabinets	13 days, $6,000, Cabinet Sub
Install Cabinets	3 days, $1,580, Cabinet Sub
Install Bathroom Accessories	½ day, $545, Builder
Order Appliances	1 day, $0, Builder
Deliver and Set Appliances	1 day, $3,735, Builder
Dishwasher	
Stove	
Compactor	
Disposal	
Range Hood	
Finish Utilities	
Plumbing Trim-out	1 day, $2,314, Plumber
HVAC Trim-out	1 day, $6,419, AC Sub
Deliver HVAC Equipment w/ above	
Set grills and Thermostat w/ above	
Start Up and Test Equipment w/ above	
Electrical Trim-out	1 day, $4,137, Electrician
Install Fixtures, Devices w/ above	
Order Light Fixtures w/ above	
Deliver Light Fixtures w/ above	
Install Light Fixtures w/ above	
Hook Up Main Power w/ above	
Check and Test System w/ above	
Final Electrical Inspection	½ day, $0, Electrician

The plumber, AC subcontractor, and electrician will all complete the balance of their work during the finish phase. We will use the summary activity to track their work. The final electrical inspection will be checked to make sure the permanent power will be available for completion.

Finish Phase Diagram

Let's construct the diagram for the finish phase. One possible diagram is shown in Figure 3.7. The words that appear in boldface refer to an activity or activities listed in Figure 3.6. The logic diagram is based on the following considerations and restraints:

1. The **sheetrock is delivered, hung, taped and finished**, and **cleaned up**. During this time, the **interior trim** is delivered and the **attic insulation** is placed after the sheetrock is taped and finished.
2. The **interior trim** and **doors** are installed before the painter begins the **interior paint**. After interior painting is completed, the **wallpaper** and **vinyl flooring** are installed. The painter can also do the **exterior painting** as weather permits. The **cabinets** are delivered and installed after the wallpaper and vinyl are in place.
3. Once the cabinets are in place, the **ceramic tile** and **clay tile** are started. At the same time, the trim subcontractor installs **hardware** on the doors, then **shelving**, and the builder will install the **bathroom accessories**. The **HVAC trim-out** also begins at this point.
4. **Plumbing trim-out** must wait until after the **ceramic tile** has been installed. **Appliances** will be delivered and set after the clay tile is finished in the kitchen area. **Electrical trim-out** follows these activities. The electrician can then hook up the air conditioning units, water heater, and the appliances.
5. The **imported marble** should be delivered by the time the electricians are finished. We have placed a soft restraint in the diagram at this point to delay the installation of the marble. Trying to protect this material is a priority, so the marble will be one of the last items to be installed.
6. The last finish phase activities are clean up in preparation for the **carpet** installation and a final touch-up on the paint. An **electrical inspection** is needed to get the electrical meter set and permanent power turned on. The builder will then do a **final inspection**.

Finish Site, Additional Special Items, Owner Activities, and Project Close-out Phases

Our master activity list includes some special items and additional project close-out activities. These items will be identified or omitted from our master project list as shown in Figure 3.8.

Because these few activities span the entire project, we will not prepare a separate diagram for them. Instead, these activities will be incorporated into the final complete diagram of the house. The owner selections listed above are simply examples; other selections and owner activities could be required, depending on project circumstances and builder preferences.

Figure 3.7: Partial Finish Phase Diagram

Phase/Activity List	Resp	Duration	Start	End
Finish Phase		29.00 d	Oct/23/96	Dec/04/96
Order Sheetrock	Builder	1.00 d	Oct/23/96	Oct/23/96
Deliver Sheetrock	Builder	1.00 d	Oct/24/96	Oct/24/96
Hang Sheetrock	Drywall Sub	2.00 d	Oct/25/96	Oct/28/96
Insulate Attic	Insulator	1.00 d	Nov/04/96	Nov/04/96
Tape & Finish Sheetrock	Drywall Sub	4.00 d	Oct/29/96	Nov/01/96
Clean up Shtrock Waste	Drywall Sub	1.00 d	Nov/04/96	Nov/04/96
Install Interior Doors	Trim Sub	1.00 d	Nov/05/96	Nov/05/96
Install Interior Trim	Trim Sub	2.00 d	Nov/06/96	Nov/07/96
Order & Deliver Int Trim	Builder	2.00 d	Oct/31/96	Nov/01/96
Paint Exterior	Painter	4.00 d	Oct/23/96	Oct/28/96
Install Vinyl Tile	Floor Sub	1.00 d	Nov/15/96	Nov/15/96
Paint Interior	Painter	4.00 d	Nov/08/96	Nov/14/96
Order Cabinets	Builder	1.00 d	Oct/29/96	Oct/29/96
Build & Deliver Cabinets	Cab Sub	13.00 d	Oct/30/96	Nov/18/96
Install Cabinets	Cab Sub	3.00 d	Nov/21/96	Nov/25/96
Order Appliances	Builder	1.00 d	Oct/30/96	Oct/30/96
Deliver & Set Appliances	Builder	1.00 d	Dec/04/96	Dec/04/96
Select Wallpaper	Owner	1.00 d	Oct/23/96	Oct/23/96
Order Wallpaper	Builder	1.00 d	Oct/24/96	Oct/24/96
Deliver Wallpaper	Painter	9.00 d	Oct/25/96	Nov/06/96
Hang Wallpaper	Painter	4.00 d	Nov/15/96	Nov/20/96
Install Clay Tile	Floor Sub	5.00 d	Nov/26/96	Dec/03/96
Install Ceramic Tile	Floor Sub	2.00 d	Nov/26/96	Nov/27/96
Plumbing Trim-out	Plumber	1.00 d	Nov/29/96	Nov/29/96
HVAC Trim-out	AC Sub	1.00 d	Nov/26/96	Nov/26/96
Install Door Hardware	Trim Sub	2.00 d	Nov/15/96	Nov/18/96
Install Shelving & Rods	Trim Sub	2.00 d	Nov/19/96	Nov/20/96
Install Bathroom Access	Builder	0.50 d	Nov/21/96	Nov/21/96
Select Entry Marble	Owner	1.00 d	Oct/23/96	Oct/23/96
Order Marble	Builder	1.00 d	Oct/24/96	Oct/24/96
Deliver Marble	Floor Sub	25.00 d	Oct/25/96	Dec/02/96
Install Marble	Floor Sub	2.00 d	Dec/03/96	Dec/04/96
Install Carpet	Floor Sub	2.00 d	Dec/03/96	Dec/04/96
Electrical Trim-out	Electrician	1.00 d	Dec/02/96	Dec/02/96
Final Electrical Inspection	Electrician	0.50 d	Dec/03/96	Dec/03/96

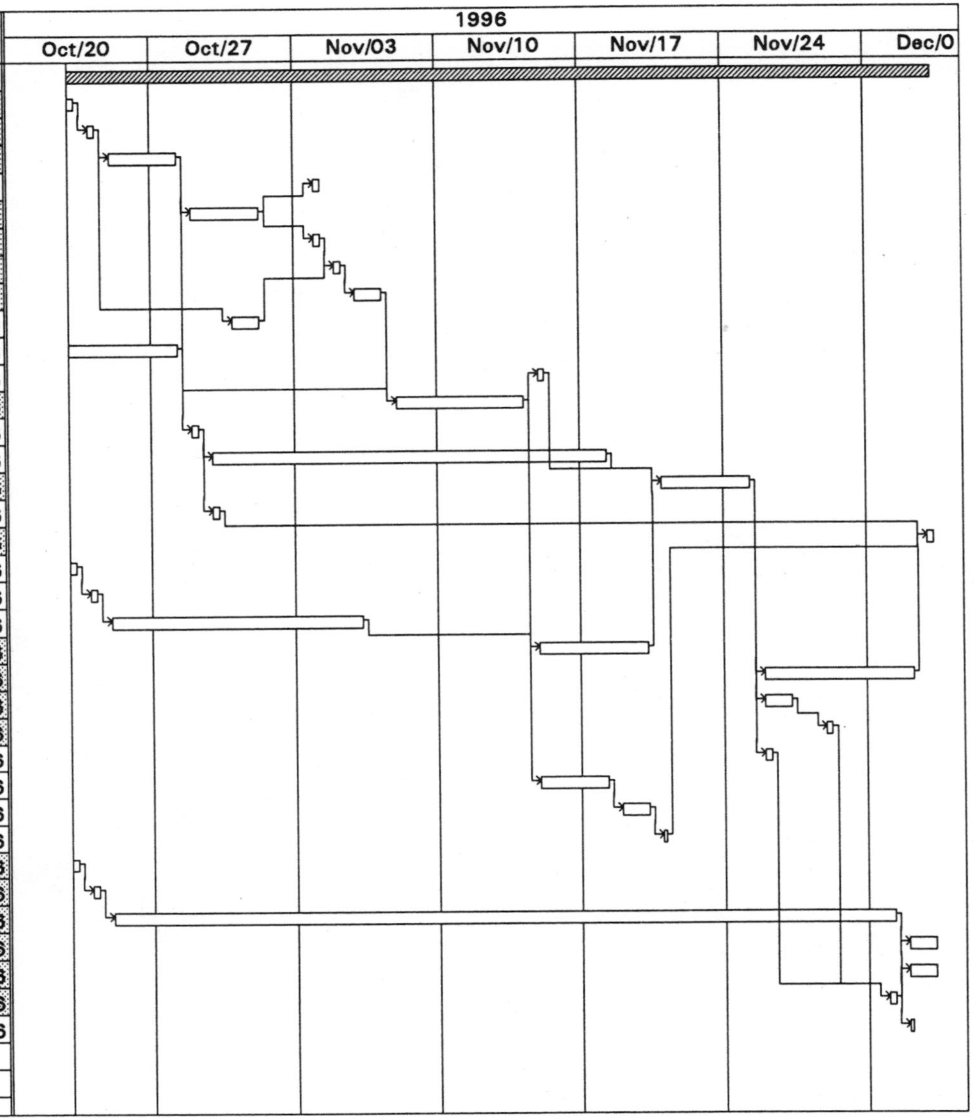

Shaded activities are on the critical path.

Figure 3.8: Plan for Special Items, Finish Sitework, and Project Close-out

Activity	Duration, Cost, Responsibility
Owner Activities	
Owner Select Brick	1 day, $0, Owner
Owner Select Wallpaper	1 day, $0, Owner
Owner Select Whirlpool Tub	1 day, $0, Owner
Owner Select Entry Marble	1 day, $0, Owner
Custom Items	
Install Outside Decks	Not Required
Install Hot Tubs	Not Required
Install Alarm Systems	Not Required
Install Telephone Wiring	Electrician Included
Install TV Wiring	Electrician Included
Finish Sitework	
Grade Driveway & Walks	½ day, $250, Site Sub
Install Concrete Driveway	3 days, $4,996, Site Sub
Form and Place Sidewalks	1 day, $854, Site Sub
Site Landscaping	3 days, $2,500, Landscaper
Project Close-out Phase	
Remove Temp. Water Connection	Plumber Included
Remove Temp. Power Connection	Electrician Included
Paint Touch-up	½ day, $0, Painter
Clean House	½ day, $500, Builder
Final Inspection	½ day, $0, Builder
Complete Punch List Work	½ day, $0, Builder

COMPLETE SCHEDULE DIAGRAM FOR HOUSE

When all of the activities have been listed, assigned to phases, and assigned their duration, cost, and responsibility, we can create a complete network diagram. Figure 3.9 presents a time-scaled, activity-on-node diagram for our sample house project.

Figure 3.9 combines the separate partial diagrams created in this chapter into one diagram. Each phase contains certain activities that connect to activities in the next phase. This diagram contains all the activities we have detailed above, as well as the activities listed in Figure 3.8.

Once the diagrams have been drawn, the builder can complete the network calculations by conducting the forward and backward passes discussed in Chapter 2. The information from doing the network calculations for the AON diagram appears in Figure 3.10. This information was needed to create the time-scaled diagrams.

In reviewing the information contained in Figure 3.10, the critical activities in this project are identified by looking at the float. If the float was zero, then the activity is critical. The critical path of a project typically includes the following activities: slab activities, framing and roofing, sheetrock, interior trim, painting, and some finishes to the final activities.

The shingles are installed between Workday 32 and Workday 36. We have followed the rule of thumb that the dry-in should occur near the middle of the project's time frame. The

Figure 3.9: Schedule for the Complete House

Phase/Activity List	Resp	Duration	Start	End
Project Startup Phase		11.00 d	Sep/09/96	Sep/23/96
Obtain Permit	Builder	3.00 d	Sep/09/96	Sep/11/96
Obtain Temporary Power	Electrician	2.00 d	Sep/12/96	Sep/13/96
Obtain Temporary Water	Plumber	2.00 d	Sep/12/96	Sep/13/96
Surv Locate Bldg Corners	Surveyor	3.00 d	Sep/19/96	Sep/23/96
Sitework Phase		5.00 d	Sep/12/96	Sep/18/96
Site Grading	Site Sub	5.00 d	Sep/12/96	Sep/18/96
Slab-on-Grade Phase		12.00 d	Sep/24/96	Oct/09/96
Lay Out & Set Batterboard	Builder	1.00 d	Sep/24/96	Sep/24/96
Excavate Turndown Edge	Builder	0.50 d	Sep/25/96	Sep/25/96
Set Edge Forms	Builder	1.50 d	Sep/25/96	Sep/26/96
HVAC Rough-in	AC Sub	0.50 d	Sep/25/96	Sep/25/96
Radon System Rough-in	Radon Sub	0.50 d	Sep/25/96	Sep/25/96
Spread Gravel	Builder	0.50 d	Sep/30/96	Sep/30/96
Plumbing Rough-in	Plumber	3.00 d	Sep/25/96	Sep/27/96
Termite Treatment	Termite	0.50 d	Sep/30/96	Sep/30/96
Set Grade Stakes	Builder	0.50 d	Oct/01/96	Oct/01/96
Place Poly and WWM	Builder	0.50 d	Oct/01/96	Oct/01/96
Inspect Slab	Inspector	0.50 d	Oct/01/96	Oct/01/96
Place & Finish Concrete	Builder	1.00 d	Oct/02/96	Oct/02/96
Place Garage/Storage Slab	Builder	3.00 d	Oct/03/96	Oct/07/96
Place Concrete Patio	Builder	2.00 d	Oct/08/96	Oct/09/96
Structure Phase		26.50 d	Sep/23/96	Oct/30/96
Order Framing Material	Builder	2.00 d	Sep/27/96	Sep/30/96
Deliver Framing Material	Builder	1.00 d	Oct/01/96	Oct/01/96
Wall and Sheathing	Framing	3.00 d	Oct/03/96	Oct/07/96
Roof Framing & Decking	Framing	5.00 d	Oct/10/96	Oct/17/96
Select Whirlpool Tub	Owner	1.00 d	Sep/23/96	Sep/23/96
Order & Deliver Tubs	Builder	4.00 d	Sep/24/96	Sep/27/96
Plumbing Top-out	Plumber	4.00 d	Oct/18/96	Oct/23/96
Plumbing Test & Insp	Plumber	0.50 d	Oct/24/96	Oct/24/96
HVAC Rough-in	AC Sub	3.00 d	Oct/18/96	Oct/22/96
Radon System Top-out	Radon Sub	1.00 d	Oct/25/96	Oct/25/96
Install Fireplace	Fire Sub	1.00 d	Oct/08/96	Oct/08/96
Electrical Rough-in	Electrician	2.00 d	Oct/23/96	Oct/24/96
Electrical Inspection	Electrician	0.50 d	Oct/25/96	Oct/25/96

1996: Sep, Oct, Nov, De

Shaded activities are on the critical path.

Figure 3.9: Schedule for the Complete House, Continued

Phase/Activity List	Resp	Duration	Start	End
Install Deadwood and Blk	Framing	0.50 d	Oct/18/96	Oct/18/96
Check Plumb & Square	Builder	0.50 d	Oct/18/96	Oct/18/96
Framing Inspection	Builder	0.50 d	Oct/18/96	Oct/18/96
Clean up All Material	Builder	0.50 d	Oct/28/96	Oct/28/96
City Inspection	Inspector	0.50 d	Oct/28/96	Oct/28/96
Wall Insulation	Insulator	1.00 d	Oct/29/96	Oct/30/96
Structure Dry-in Phase		27.50 d	Sep/23/96	Oct/31/96
Install Fascia and Soffits	Framing	4.00 d	Oct/18/96	Oct/24/96
Install Shingles	Roofer	3.00 d	Oct/24/96	Oct/29/96
Order Ext Doors & Win	Builder	10.00 d	Sep/23/96	Oct/04/96
Deliver Ext Doors & Win	Builder	1.00 d	Oct/07/96	Oct/07/96
Install Ext Doors & Win	Framing	4.00 d	Oct/18/96	Oct/24/96
Install Overhead Door	Framing	0.50 d	Oct/24/96	Oct/24/96
Install Siding	Framing	0.50 d	Oct/25/96	Oct/25/96
Select Brick	Owner	1.00 d	Sep/24/96	Sep/24/96
Order Brick	Builder	1.00 d	Sep/25/96	Sep/25/96
Deliver Brick	Builder	15.00 d	Sep/26/96	Oct/17/96
Install Brick	Mason	4.00 d	Oct/25/96	Oct/31/96
Deliver Synthetic Stucco	Stucco Sub	0.50 d	Oct/15/96	Oct/15/96
Set Panels and Trim	Stucco Sub	0.50 d	Oct/24/96	Oct/24/96
Apply Finish	Stucco Sub	3.00 d	Oct/25/96	Oct/29/96
Install Chimney Cap	Fire Sub	0.50 d	Oct/30/96	Oct/30/96
Finish Phase		52.50 d	Sep/24/96	Dec/11/96
Order Sheetrock	Builder	1.00 d	Oct/17/96	Oct/17/96
Deliver Sheetrock	Builder	1.00 d	Oct/18/96	Oct/18/96
Hang Sheetrock	Drywall Sub	2.00 d	Oct/30/96	Nov/01/96
Insulate Attic	Insulator	1.00 d	Nov/07/96	Nov/08/96
Tape & Finish Sheetrock	Drywall Sub	4.00 d	Nov/01/96	Nov/07/96
Clean up Sheetrock Waste	Drywall Sub	1.00 d	Nov/07/96	Nov/08/96
Install Interior Doors	Trim Sub	1.00 d	Nov/08/96	Nov/12/96
Install Interior Trim	Trim Sub	2.00 d	Nov/12/96	Nov/14/96
Order & Deliver Int Trim	Builder	2.00 d	Oct/21/96	Oct/22/96
Paint Exterior	Painter	4.00 d	Oct/31/96	Nov/06/96
Install Vinyl Tile	Floor Sub	1.00 d	Nov/20/96	Nov/21/96
Paint Interior	Painter	4.00 d	Nov/14/96	Nov/20/96
Order Cabinets	Builder	1.00 d	Oct/11/96	Oct/11/96

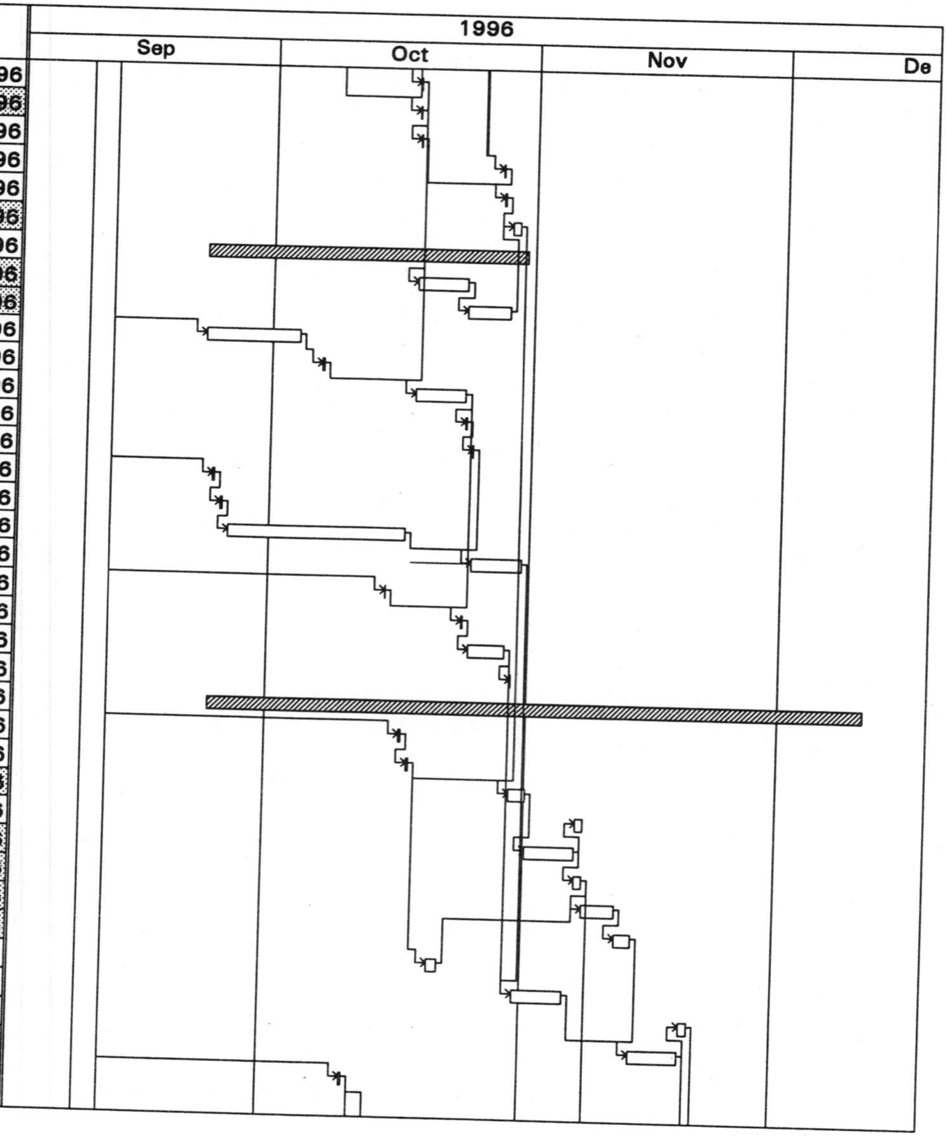

Shaded activities are on the critical path.

Figure 3.9: Schedule for the Complete House, Continued

Phase/Activity List	Resp	Duration	Start	End
Build & Deliver Cabinets	Cab Sub	13.00 d	Oct/15/96	Oct/31/96
Install Cabinets	Cab Sub	3.00 d	Nov/26/96	Dec/03/96
Order Appliances	Builder	1.00 d	Oct/15/96	Oct/15/96
Deliver & Set Appliances	Builder	1.00 d	Dec/10/96	Dec/11/96
Select Wallpaper	Owner	1.00 d	Oct/23/96	Oct/23/96
Order Wallpaper	Builder	1.00 d	Oct/24/96	Oct/24/96
Deliver Wallpaper	Painter	9.00 d	Oct/25/96	Nov/06/96
Hang Wallpaper	Painter	4.00 d	Nov/20/96	Nov/26/96
Install Clay Tile	Floor Sub	5.00 d	Dec/03/96	Dec/10/96
Install Ceramic Tile	Floor Sub	2.00 d	Dec/03/96	Dec/05/96
Plumbing Trim-out	Plumber	1.00 d	Dec/05/96	Dec/06/96
HVAC Trim-out	AC Sub	1.00 d	Dec/03/96	Dec/04/96
Install Door Hardware	Trim Sub	2.00 d	Nov/20/96	Nov/22/96
Install Shelving & Rods	Trim Sub	2.00 d	Nov/22/96	Nov/26/96
Install Bathroom Access	Builder	0.50 d	Nov/26/96	Nov/26/96
Select Entry Marble	Owner	1.00 d	Sep/24/96	Sep/24/96
Order Marble	Builder	1.00 d	Sep/25/96	Sep/25/96
Deliver Marble	Floor Sub	40.00 d	Sep/26/96	Nov/22/96
Install Marble	Floor Sub	2.00 d	Dec/09/96	Dec/11/96
Install Carpet	Floor Sub	2.00 d	Dec/09/96	Dec/11/96
Electrical Trim-out	Electrician	1.00 d	Dec/06/96	Dec/09/96
Final Electrical Inspection	Electrician	0.50 d	Dec/09/96	Dec/09/96
Finish Sitework Phase		7.50 d	Nov/08/96	Nov/20/96
Grade Driveway & Walks	Site Sub	0.50 d	Nov/08/96	Nov/08/96
Install Concrete Driveway	Site Sub	3.00 d	Nov/12/96	Nov/14/96
Form & Place Sidewalks	Site Sub	1.00 d	Nov/15/96	Nov/15/96
Landscaping	Landscaper	3.00 d	Nov/18/96	Nov/20/96
Project Close-out Phase		2.00 d	Dec/11/96	Dec/13/96
Paint Touch-up	Painter	0.50 d	Dec/11/96	Dec/11/96
Clean House	Builder	0.50 d	Dec/12/96	Dec/12/96
Complete Punch List	Builder	0.50 d	Dec/12/96	Dec/12/96
Final Inspection	Builder	0.50 d	Dec/13/96	Dec/13/96

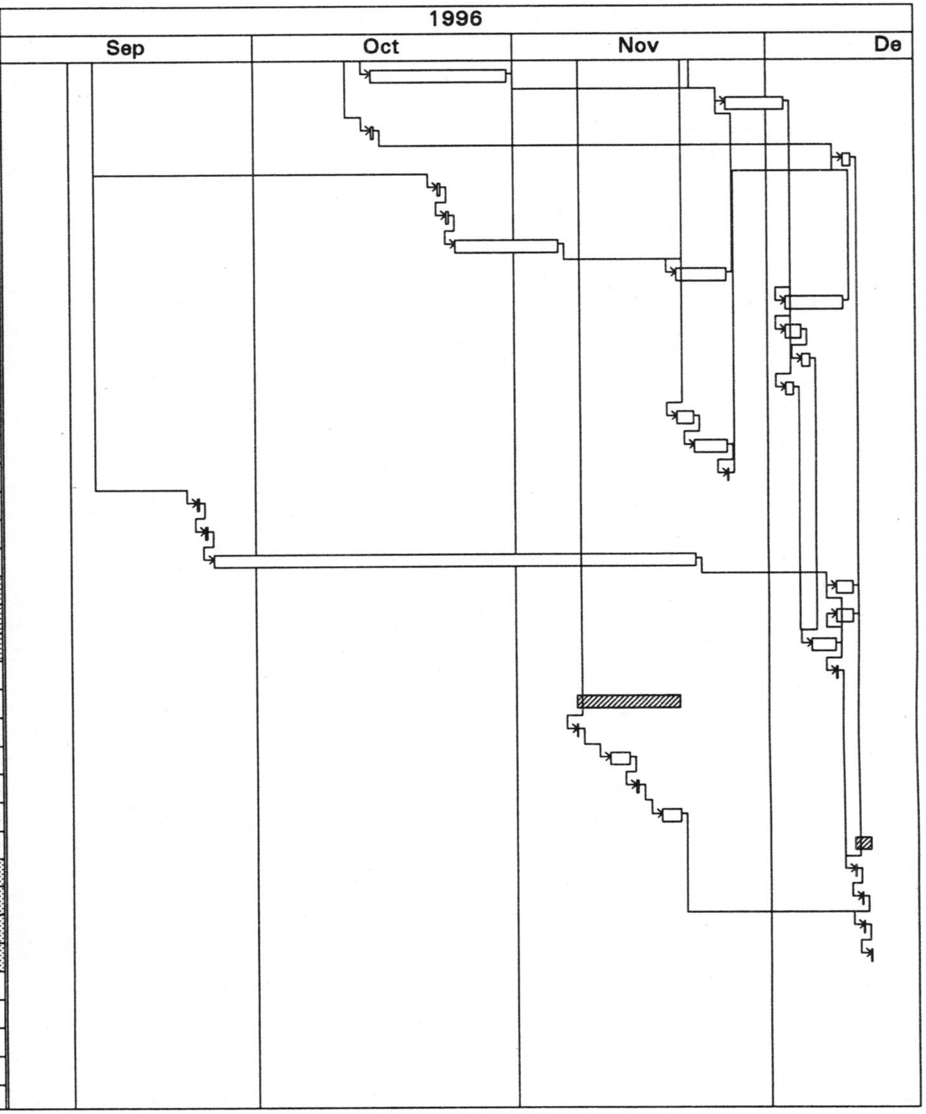

1996: Sep | Oct | Nov | De

Shaded activities are on the critical path.

Figure 3.10: Network Calculations for the Sample House

Network Calculations
Activity-on-Node Diagram for Complete House
Project Start Date = 9/9/96
Project Finish Date = 12/13/96

Holidays 10/14 for Columbus Day
11/11 for Veteran's Day
11/28 & 29 for Thanksgiving

			WORK DAY NUMBERS				CALENDAR DATES				
Phase/Activity List	Resp	Duration	Early Start	Early Finish	Late Start	Late Finish	Early Start	Early Finish	Late Start	Late Finish	Float
Project Startup Phase											
Obtain Permit	Builder	3.00 d	0	3	0	3	9/9/96	9/11/96	9/9/96	9/11/96	0
Obtain Temporary Power	Electrician	2.00 d	3	5	9	11	9/12/96	9/13/96	9/20/96	9/23/96	6
Obtain Temporary Water	Plumber	2.00 d	3	5	10	12	9/12/96	9/13/96	9/23/96	9/24/96	7
Surv Locate Bldg Corners	Surveyor	3.00 d	8	11	8	11	9/19/96	9/23/96	9/19/96	9/23/96	0
Sitework Phase											
Site Grading	Site Sub	5.00 d	3	8	3	8	9/12/96	9/18/96	9/12/96	9/18/96	0
Slab-on-Grade Phase											
Lay Out & Set Batterboards	Builder	1.00 d	11	12	11	12	9/24/96	9/24/96	9/24/96	9/24/96	0
Excavate Turndown Edge	Builder	0.50 d	12	12.5	13.5	14	9/25/96	9/25/96	9/26/96	9/26/96	1.5
Set Edge Forms	Builder	1.50 d	12	13.5	15.5	15	9/25/96	9/26/96	9/26/96	9/27/96	3.5
HVAC Rough-in	AC Sub	0.50 d	12	12.5	14.5	15	9/25/96	9/25/96	9/27/96	9/27/96	2.5
Radon System Rough-in	Radon Sub	0.50 d	12	12.5	14.5	15	9/25/96	9/25/96	9/27/96	9/27/96	2.5
Spread Gravel	Builder	0.50 d	15	15.5	15	16	9/30/96	9/30/96	9/30/96	9/30/96	0
Plumbing Rough-in	Plumber	3.00 d	12	15	12	15	9/25/96	9/27/96	9/25/96	9/27/96	0
Termite Treatment	Termite	0.50 d	15	15.5	15	16	9/30/96	9/30/96	9/30/96	9/30/96	0
Set Grade Stakes	Builder	0.50 d	16	16.5	16	17	10/1/96	10/1/96	10/1/96	10/1/96	0
Place Poly and WWM	Builder	0.50 d	16	16.5	16	17	10/1/96	10/1/96	10/1/96	10/1/96	0
Inspect Slab	Inspector	0.50 d	16	16.5	16	17	10/1/96	10/1/96	10/1/96	10/1/96	0
Place & Finish Concrete	Builder	1.00 d	17	18	17	18	10/2/96	10/2/96	10/2/96	10/2/96	0
Place Garage/Storage Slab	Builder	3.00 d	18	21	18	21	10/3/96	10/7/96	10/3/96	10/7/96	0
Place Concrete Patio	Builder	2.00 d	21	23	21	23	10/8/96	10/9/96	10/8/96	10/9/96	0
Structure Phase											
Order Framing Material	Builder	2.00 d	14	16	17	19	9/27/96	9/30/96	10/2/96	10/3/96	3
Deliver Framing Material	Builder	1.00 d	16	17	19	20	10/1/96	10/1/96	10/4/96	10/4/96	3
Wall and Sheathing	Framing	3.00 d	18	21	20	23	10/3/96	10/7/96	10/7/96	10/9/96	2
Roof Framing & Decking	Framing	5.00 d	23	28	23	28	10/10/96	10/17/96	10/10/96	10/17/96	0
Select Whirlpool Tub	Owner	1.00 d	10	11	15	16	9/23/96	9/23/96	9/30/96	9/30/96	5
Order & Deliver Tubs	Builder	4.00 d	11	15	16	20	9/24/96	9/27/96	10/1/96	10/4/96	5
Plumbing Top-out	Plumber	4.00 d	28	32	29	33	10/18/96	10/23/96	10/21/96	10/24/96	1
Plumbing Test & Insp	Plumber	0.50 d	32	32.5	33.5	34	10/24/96	10/24/96	10/25/96	10/25/96	1.5
HVAC Rough-in	AC Sub	3.00 d	28	31	29	32	10/18/96	10/22/96	10/18/96	10/23/96	1
Radon System Top-out	Radon Sub	1.00 d	33	34	34	35	10/25/96	10/25/96	10/25/96	10/28/96	1
Install Fireplace	Fire Sub	1.00 d	21	22	27	28	10/8/96	10/8/96	10/17/96	10/17/96	6
Electrical Rough-in	Electrician	2.00 d	31	33	32	34	10/23/96	10/24/96	10/23/96	10/25/96	1

Figure 3.10: Network Calculations for the Sample House, Continued

Network Calculations
Activity-on-Node Diagram for Complete House
Project Start Date = 9/9/96
Project Finish Date = 12/13/96

Holidays: 10/14 for Columbus Day; 11/11 for Veteran's Day; 11/28 & 29 for Thanksgiving

			WORK DAY NUMBERS				CALENDAR DATES				
Phase/Activity List	Resp	Duration	Early Start	Early Finish	Late Start	Late Finish	Early Start	Early Finish	Late Start	Late Finish	Float
Electrical Inspection	Electrician	0.50 d	33	33.5	34.5	35	10/25/96	10/25/96	10/28/96	10/28/96	1.5
Install Deadwood and Blk	Framing	0.50 d	28	28.5	34.5	35	10/18/96	10/18/96	10/28/96	10/28/96	6.5
Check Plumb & Square	Builder	0.50 d	28	38.5	28	29	10/18/96	10/18/96	10/18/96	10/18/96	0
Framing Inspection	Builder	0.50 d	28	28.5	34.5	35	10/18/96	10/18/96	10/28/96	10/28/96	6.5
Clean up All Material	Builder	0.50 d	34	34.5	34.5	35	10/28/96	10/28/96	10/28/96	10/28/96	0.5
City Inspection	Inspector	0.50 d	34	34.5	35.5	36	10/28/96	10/28/96	10/29/96	10/29/96	1.5
Wall Insulation	Insulator	1.00 d	35	36	35	37	10/29/96	10/30/96	10/29/96	10/30/96	0
Structure Dry-in Phase											
Install Fascia and Soffits	Framing	4.00 d	28	32	28	33	10/18/96	10/24/96	10/18/96	10/24/96	0
Install Shingles	Roofer	3.00 d	32	35	32	36	10/24/96	10/29/96	10/24/96	10/29/96	0
Order Ext Doors & Win	Builder	10.00 d	10	20	23	33	9/23/96	10/4/96	10/9/96	10/24/96	13
Deliver Ext Doors & Win	Builder	1.00 d	20	21	34	35	10/7/96	10/7/96	10/24/96	10/25/96	14
Install Ext Doors & Win	Framing	4.00 d	28	32	34	38	10/18/96	10/24/96	10/25/96	10/31/96	6
Install Overhead Door	Framing	0.50 d	32	32.5	37.5	38	10/24/96	10/24/96	10/31/96	10/31/96	5.5
Install Siding	Framing	0.50 d	33	33.5	38.5	39	10/25/96	10/25/96	11/1/96	11/1/96	5.5
Select Brick	Owner	1.00 d	11	12	22	23	9/24/96	9/24/96	10/8/96	10/9/96	11
Order Brick	Builder	1.00 d	12	13	23	24	9/25/96	9/25/96	10/9/96	10/10/96	11
Deliver Brick	Builder	15.00 d	13	28	24	39	9/26/96	10/17/96	10/10/96	11/1/96	11
Install Brick	Mason	4.00 d	33	37	39	43	10/25/96	10/31/96	11/1/96	11/7/96	6
Deliver Synthetic Stucco	Stucco Sub	0.50 d	25	25.5	38.5	39	10/15/96	10/15/96	11/1/96	11/1/96	13.5
Set Panels and Trim	Stucco Sub	0.50 d	32	32.5	39.5	40	10/24/96	10/24/96	11/4/96	11/4/96	7.5
Apply Finish	Stucco Sub	3.00 d	33	36	40	43	10/25/96	10/29/96	11/4/96	11/7/96	7
Install Chimney Cap	Fire Sub	0.50 d	36	36.5	38	39	10/30/96	10/30/96	11/1/96	11/1/96	2
Finish Phase											
Order Sheetrock	Builder	1.00 d	27	28	35	36	10/17/96	10/17/96	10/28/96	10/29/96	8
Deliver Sheetrock	Builder	1.00 d	28	29	36	37	10/18/96	10/18/96	10/29/96	10/30/96	8
Hang Sheetrock	Drywall Sub	2.00 d	36	38	36	39	10/30/96	11/1/96	10/30/96	11/1/96	0
Insulate Attic	Insulator	1.00 d	42	43	63	64	11/7/96	11/8/96	12/10/96	12/11/96	21
Tape & Finish Sheetrock	Drywall Sub	4.00 d	38	42	38	43	11/1/96	11/7/96	11/1/96	11/7/96	0
Clean up Sheetrock Waste	Drywall Sub	1.00 d	42	43	42	44	11/7/96	11/8/96	11/7/96	11/8/96	0
Install Interior Doors	Trim Sub	1.00 d	43	44	43	45	11/8/96	11/12/96	11/8/96	11/12/96	0
Install Interior Trim	Trim Sub	2.00 d	44	46	44	47	11/12/96	11/14/96	11/12/96	11/14/96	0
Order & Deliver Int Trim	Builder	2.00 d	29	31	43	44	10/21/96	10/22/96	11/6/96	11/8/96	14
Paint Exterior	Painter	4.00 d	37	41	43	47	10/31/96	11/6/96	11/7/96	11/14/96	6
Install Vinyl Tile	Floor Sub	1.00 d	50	51	54	55	11/20/96	11/21/96	11/25/96	11/26/96	4

Figure 3.10: Network Calculations for the Sample House, Continued

Network Calculations
Activity-on-Node Diagram for Complete House
Project Start Date = 9/9/96
Project Finish Date = 12/13/96

Holidays: 10/14 for Columbus Day; 11/11 for Veteran's Day; 11/28 & 29 for Thanksgiving

			WORK DAY NUMBERS				CALENDAR DATES				
Phase/Activity List	Resp	Duration	Early Start	Early Finish	Late Start	Late Finish	Early Start	Early Finish	Late Start	Late Finish	Float
Paint Interior	Painter	4.00 d	46	50	46	51	11/14/96	11/20/96	11/14/96	11/20/96	0
Order Cabinets	Builder	1.00 d	24	25	41	42	10/11/96	10/11/96	11/5/96	11/6/96	17
Build & Deliver Cabinets	Cab Sub	13.00 d	25	38	42	55	10/15/96	10/31/96	11/6/96	11/26/96	17
Install Cabinets	Cab Sub	3.00 d	54	57	54	58	11/26/96	12/3/96	11/26/96	12/3/96	0
Order Appliances	Builder	1.00 d	25	26	62	63	10/15/96	10/15/96	12/9/96	12/10/96	37
Deliver & Set Appliances	Builder	1.00 d	62	63	62	64	12/10/96	12/11/96	12/10/96	12/11/96	0
Select Wallpaper	Owner	1.00 d	31	32	40	41	10/23/96	10/23/96	11/4/96	11/5/96	9
Order Wallpaper	Builder	1.00 d	32	33	41	42	10/24/96	10/24/96	11/5/96	11/6/96	9
Deliver Wallpaper	Painter	9.00 d	33	42	42	51	10/25/96	11/6/96	11/6/96	11/20/96	9
Hang Wallpaper	Painter	4.00 d	50	54	50	55	11/20/96	11/26/96	11/20/96	11/26/96	0
Install Clay Tile	Floor Sub	5.00 d	57	62	57	63	12/3/96	12/10/96	12/3/96	12/10/96	0
Install Ceramic Tile	Floor Sub	2.00 d	57	59	57	60	12/3/96	12/5/96	12/3/96	12/5/96	0
Plumbing Trim-out	Plumber	1.00 d	59	60	59	61	12/5/96	12/6/96	12/5/96	12/6/96	0
HVAC Trim-out	AC Sub	1.00 d	57	58	60	61	12/3/96	12/4/96	12/5/96	12/6/96	3
Install Door Hardware	Trim Sub	2.00 d	50	52	58	60	11/20/96	11/22/96	12/4/96	12/5/96	8
Install Shelving & Rods	Trim Sub	2.00 d	52	54	60	62	11/22/96	11/26/96	12/6/96	12/9/96	8
Install Bathroom Access	Builder	0.50 d	54	54.5	62.5	63	11/26/96	11/26/96	12/10/96	12/10/96	8.5
Select Entry Marble	Owner	1.00 d	11	12	20	21	9/24/96	9/24/96	10/4/96	10/7/96	9
Order Marble	Builder	1.00 d	12	13	21	22	9/25/96	9/25/96	10/7/96	10/8/96	9
Deliver Marble	Floor Sub	40.00 d	13	53	42	62	9/26/96	11/22/96	10/8/96	12/9/96	29
Install Marble	Floor Sub	2.00 d	61	63	61	64	12/9/96	12/11/96	12/9/96	12/11/96	0
Install Carpet	Floor Sub	2.00 d	61	63	61	64	12/9/96	12/11/96	12/9/96	12/11/96	0
Electrical Trim-out	Electrician	1.00 d	60	61	60	62	12/6/96	12/9/96	12/6/96	12/9/96	0
Final Electrical Inspection	Electrician	0.50 d	61	61.5	63.5	64	12/9/96	12/9/96	12/11/96	12/11/96	2.5
Finish Sitework Phase											
Grade Driveway & Walks	Site Sub	0.50 d	41	41.5	57.5	58	11/8/96	11/8/96	12/3/96	12/3/96	16.5
Install Concrete Driveway	Site Sub	3.00 d	44	47	58	61	11/12/96	11/14/96	12/3/96	12/6/96	14
Form & Place Sidewalks	Site Sub	1.00 d	47	48	61	62	11/15/96	11/15/96	12/6/96	12/9/96	14
Landscaping	Landscaper	3.00 d	48	51	62	65	11/18/96	11/20/96	12/9/96	12/12/96	14
Project Close-out Phase											
Paint Touch-up	Painter	0.50 d	63	63.5	63	64	12/11/96	12/11/96	12/11/96	12/11/96	0
Clean House	Builder	0.50 d	64	64.5	64	65	12/12/96	12/12/96	12/12/96	12/12/96	0
Complete Punch List	Builder	0.50 d	65	65.5	65	65	12/12/96	12/12/96	12/12/96	12/12/96	0
Final Inspection	Builder	0.50 d	66	66.5	66	66	12/13/96	12/13/96	12/13/96	12/13/96	0

calculated halfway point would be thirty-three workdays (66 days ÷ 2). It appears that our schedule is balanced and will provide us with a sound planning tool.

We should now be able to proceed with the scheduling process, communicating and gathering information about our project as it progresses.

SUMMARY POINTS AND TIPS

- Use your master list of activities as a planning checklist so that you don't overlook items that need to be scheduled. Don't forget customer or owner activities. Remember, the list is only a guide; you can change the level of detail as the project requires. Also, for convenience, you can combine related small-duration activities into larger, more general categories to reduce the number of activities listed in the schedule.
- Estimate durations for each task by calculating the time required for a set number of workers to complete their work. Adjust the durations to allow for anticipated project changes such as having to reduce the number of workers assigned to a task.
- Note who is responsible for each activity. Only one crew or subcontractor should be responsible for completing an activity.
- The estimate is the basis for the schedule. Using sound estimating procedures, assign resources and costs to all activities, including subcontractor activities. A scheduling system that incorporates information on percent of cost, amount installed, or time spent, allows you to manage cash flow by projecting the amounts due to the subcontractors at different points in time.
- Schedule inspections as separate activities. Inspections act as important checkpoints for the project and everyone should focus on trying to meet these important deadlines.
- Ordering and delivering materials to the right place at the right time is important. Schedule deliveries based on the point in time the material will be required on the project site. Using a schedule can help save costs by reducing storage time, waste, and theft, provided the builder allows the proper amount of lead time for delivery of specialty items.

Chapter 4

Using the Schedule

Builders can take advantage of several important uses for a schedule. We will discuss the following topics:

- Communication, Tracking, and Control
- Dealing With Changes
- Crashing or Expediting the Project
- Cost Forecasting
- Multiproject Scheduling

Builders can generate a great deal of valuable information from a schedule, although some may not want to spend the extra effort required to take advantage of some of these features. However, once you are comfortable with the basic uses of schedules, these additional features can further enhance your business.

COMMUNICATION, TRACKING, AND CONTROL

This chapter will focus on the *uses* of the schedule. Once the schedule has been created, some important steps still need to be performed to get the most use from your planning and scheduling. Each of the techniques discussed below can be performed using manual methods or by using a good-quality computerized scheduling program.

Communication

Communicating the proper information to the proper people, tracking the progress made on each activity, and controlling the project flow by focusing on critical activities are all essential functions of scheduling.

Communication is the key to using a schedule successfully. All subcontractors, material suppliers, and the builders' or remodelers' employees need to have essential information. Whether large or small, a building or remodeling project requires coordination of the efforts of all parties to complete it on time.

All workers and subcontractors need information from the schedule that tells them what they are to do and when they are expected to perform on the project. Providing schedule information that is customized to the individual worker or sub can be extremely helpful. Most builders will use the CPM diagram to forecast when the subs will be needed and then call the subs about a week to ten days ahead of time to notify them and get a commitment. Figure 4.1 displays the electrical subcontractor's activities for our house project. (If you are using or are considering purchasing a computer-based scheduling system, look for a program that allows you to sort and print selected activities so that you can easily create schedules for each subcontractor.)

A builder who is doing this manually might consider drawing a master bar chart that fits on an easily photocopied sheet of paper. The copies can then be highlighted to identify the subcontractors' activities. Another manual method would be to write out a separate schedule for each subcontractor. Either method will give the subcontractor the information needed. This communication is one of the key elements of a successful schedule.

The graphic part of the bar chart shows an approximate date, while specific start and end dates are listed on the left. Figure 4.1 does not show any network analysis or logic relationships. Although the builder or remodeler needs detailed information presented in a network diagram, most subcontractors basically just want to know when to perform their work. A bar chart is ideal for this kind of communication.

Subcontractors should be kept up-to-date weekly as to the project's status. They must plan their own workloads and need to adjust their schedules to manage various projects. Even when the project is progressing on schedule, subcontractors should be informed.

When you send copies of their schedules to your subcontractors, ask them to let you know if they foresee any problems. A subcontractor may have a conflict that will cause delays for the builder if adjustments are not made. By discussing the schedule and resolving any conflicts, you and your subcontractors can develop a consensus as to what is expected of each party. With improved communication, project problems can be overcome in a timely manner.

This scheduling technique becomes more valuable when you have several projects under construction at once. The schedules for all projects can be consolidated onto one sheet. We will discuss multiproject schedules in more detail later in the chapter.

Builders can use project schedules to communicate certain information to bankers, bonding or insurance companies, and the buyer or owner. Again, these parties usually are not interested in the details of the complete working schedule. A summary schedule often is all that is required for effective communication at this level. Figure 4.2 is a sample of such a summary developed from the house schedule. This summary was created by hiding all the activities under the phases that were listed in the previous chapters.

The summary schedule may be updated and redistributed at monthly intervals or only if a major variation or change occurs. This schedule is typically given to people needing a quick status report. The builder can use this as a major guideline and should know the project's overall status.

Figure 4.2 has been graphed in a monthly time frame. The start and finish dates of the phases are listed in the chart. This schedule gives some broad indicators of the project status; having a lot of detail is not the goal of the diagram. The builder should always consider who the diagrams are being created for and what their needs are in using the information.

Tracking

The second important function of the schedule is to allow the builder or remodeler to track the progress of the project. Tracking should be done in a regular, timely manner. Depending on the your preference—and on the number and types of jobs to be tracked—progress can be measured once a week, once a month, at the end of the project, or according to some other specified time frame. We recommend weekly.

Job progress can be measured by several methods. The builder or remodeler can, by visual inspection, estimate how close each activity is to completion and write down a percentage. Another technique is to determine the quantity of material installed and divide that by

Figure 4.1: Electrical Subcontractor Activities

Phase/Activity List	Resp	Duration	Start	End
Project Startup Phase		11.00 d	Sep/09/96	Sep/23/96
Obtain Temporary Power	Electrician	2.00 d	Sep/12/96	Sep/13/96
Structure Phase		26.50 d	Sep/23/96	Oct/30/96
Electrical Rough-in	Electrician	2.00 d	Oct/23/96	Oct/24/96
Electrical Inspection	Electrician	0.50 d	Oct/25/96	Oct/25/96
Finish Phase		52.50 d	Sep/24/96	Dec/11/96
Electrical Trim-out	Electrician	1.00 d	Dec/06/96	Dec/09/96
Final Electrical Inspection	Electrician	0.50 d	Dec/09/96	Dec/09/96

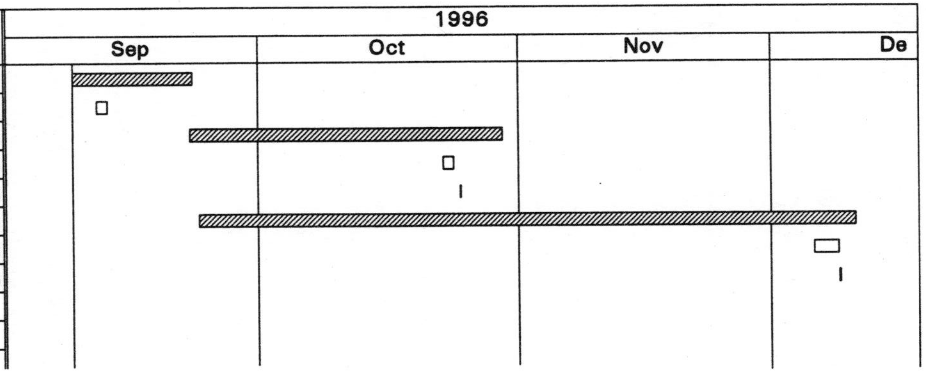

Figure 4.2: Summary Schedule

Phase/Activity List	Duration	Start	End
Project Startup Phase	11.00 d	Sep/09/96	Sep/23/96
Sitework Phase	5.00 d	Sep/12/96	Sep/18/96
Slab-on-Grade Phase	12.00 d	Sep/24/96	Oct/09/96
Structure Phase	26.50 d	Sep/23/96	Oct/30/96
Structure Dry-in Phase	27.50 d	Sep/23/96	Oct/31/96
Finish Phase	52.50 d	Sep/24/96	Dec/11/96
Finish Sitework Phase	7.50 d	Nov/08/96	Nov/20/96
Project Close-out Phase	2.00 d	Dec/11/96	Dec/13/96

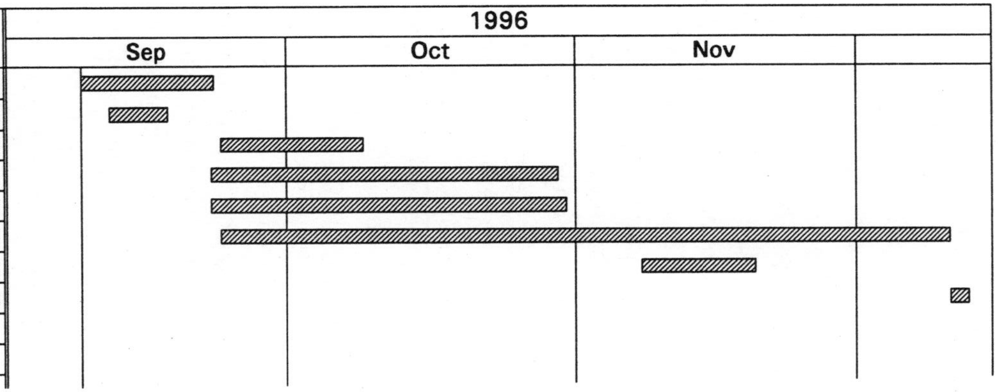

the estimated total quantity. The simplest method is the *0/100 rule*. An activity is either 0 percent complete or 100 percent complete. No credit is given for activities that are partially complete.

A version of the 0/100 rule that allows a little more fine-tuning is the *50 percent rule*. The builder or subcontractor notes the activity as either 50 percent or 100 percent complete. The advantage of using a rule like this is that it prevents argument about the status of the work. This can be an important point if subcontractors' contracts specify payments for work based on stages of completion. The subcontractor and builder should agree up front on which method will be used to measure the status of their activities.

Many builders reach an agreement with subcontractors on a set of tasks and an amount to be paid for each task when it is completed. The builder should incorporate these tasks into the schedule and link the payment with timely completion. The timely completion of the tasks is important to the builder.

A key to inspection and updating is to verify the time remaining on the activity. Someone may have worked for six days to complete an activity that was originally estimated to take five days. Recording the time remaining will uncover errors caused by optimism as well as problems occurring on the jobsite.

For example, the wall framing in our house example was originally estimated to take three days to complete. During our weekly inspection, a talk with the onsite crew reveals that the framing will take four more days. Many things could have caused the delay, such as rain, delivery of the framing material to the street rather than the house (so it has to be carried by hand), or workers who don't show. Knowing what has caused the delay may help us with our *next* project; meanwhile, however, the framing will take four more days to complete and we should reflect this fact in our updated schedule.

Once you begin to update a schedule, the level of detail can become a hindrance or a help. The more detailed the schedule, the more time it requires to update and maintain. In an update, each activity has to be dealt with in some way. For example, a small-volume builder who only updates the schedule every two weeks may not find it effective to list each half-day duration activity. On the other hand, a large-volume builder juggling multiple projects and crews (and with a staff scheduler) may prefer to monitor each schedule in great detail to prevent conflicts. The more detailed the schedule, the more time it requires to update and maintain.

As an example of tracking the work, we will update the house schedule to the end of September. We will look only at the activities that should have occurred during the first forty days of the project. The 0/50/100 percent rule will be used for updating the activities and we will verify the accuracy of the remaining durations.

The manual method would be to verify the activities that have been completed or are in progress as of September 27. Activities that are 100 percent completed have no remaining durations and can be marked as completed. Activities less than 100 percent complete should have a remaining duration. The remaining durations will be used to calculate a new project duration.

The activities are given a new start day of zero. We then do a forward and backward pass calculation to determine a new project duration. The highest number establishes the longest path and the earliest date the project will be completed.

We then convert the new workdays to calendar days by making the end of the zero day equal to the update date. For our example this would be September 27.

Figure 4.3 shows the forward and backward calculations for the garage example if it were updated on July 27. The remaining durations and network calculations are shown on the diagram. For a project of any size, this process can be time-consuming and confusing if not fully understood.

For larger diagrams such as our house example, a simple computer program can be helpful in performing the calculations. A summary activity "The House Schedule" has been added to the project in Figure 4.4. This activity represents the overall project duration. The completion date for this activity will change depending on the amount of work performed and whether that work has been completed according to its scheduled duration. The computer calculates the project's percentage of completion based on the remaining duration assigned in the update to the detailed activities.

Figure 4.3: Updated AOA Diagram for Garage

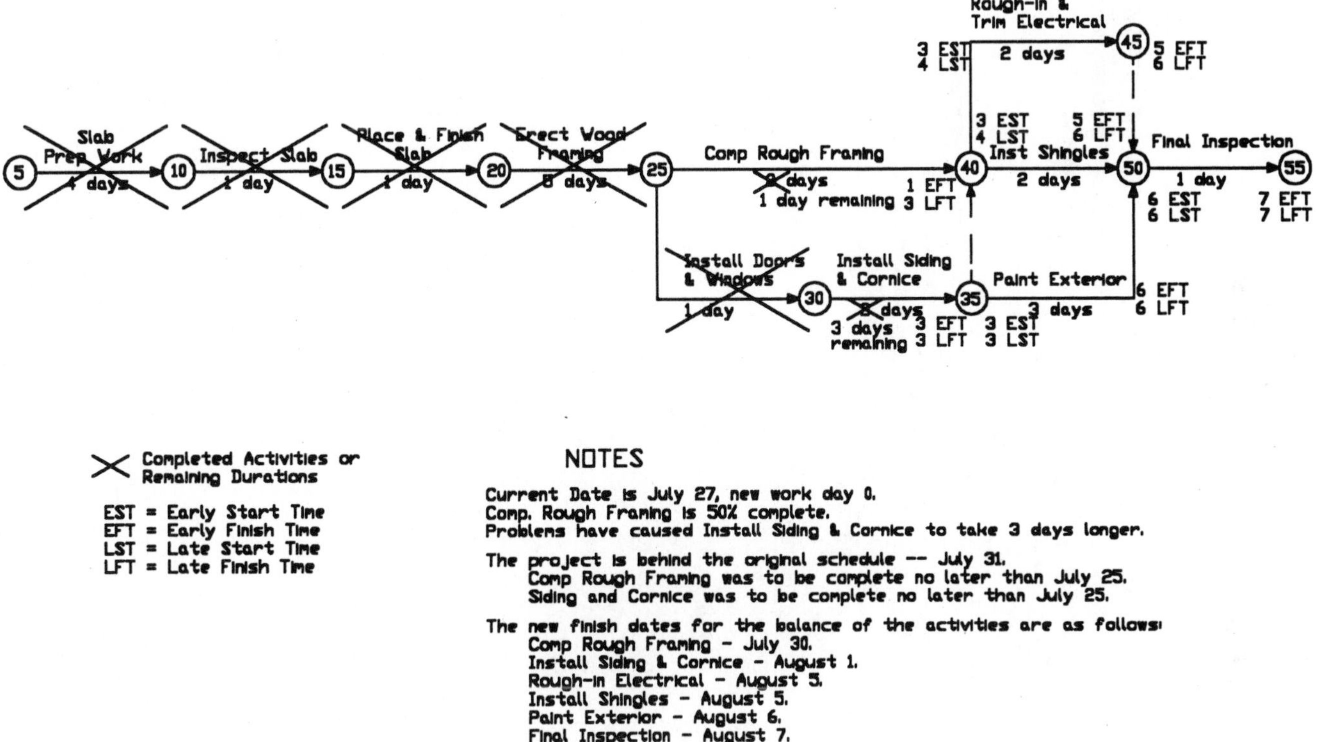

Assume we have updated our schedule after visiting the project. The vertical line that appears in the week of "Sep/29" represents the current date of September 30. If we are on schedule, all activities to the left of the line should be 100 percent complete. Overall, the project is on schedule, and the computer has calculated the summary activity as being 21 percent complete.

The structure phase and the structure dry-in phase are showing some activities as completed. These are the selections of brick and other materials listed under these phases. The materials have been ordered and the delivery dates have been established.

Figure 4.4 shows a project that is working according to the plan. What happens when a project suffers delays caused by special conditions such as weather, subcontractor problems, material shortages, or labor trouble? These conditions can be handled in two ways.

Figure 4.4: Updated Schedule

Phase/Activity List	Duratio	Remaining Duration	% Com	Start	End
The House Schedule	65.50 d	51.74 d	21	Sep/09/96	Dec/13/96
Project Startup Phase	11.00 d	0.00 d	100	Sep/09/96	Sep/23/96
Obtain Permit	3.00 d	0.00 d	100	Sep/09/96	Sep/11/96
Obtain Temporary Power	2.00 d	0.00 d	100	Sep/12/96	Sep/13/96
Obtain Temporary Water	2.00 d	0.00 d	100	Sep/12/96	Sep/13/96
Surv Locate Bldg Corners	3.00 d	0.00 d	100	Sep/19/96	Sep/23/96
Sitework Phase	5.00 d	0.00 d	100	Sep/12/96	Sep/18/96
Site Grading	5.00 d	0.00 d	100	Sep/12/96	Sep/18/96
Slab-on-Grade Phase	12.00 d	6.60 d	45	Sep/24/96	Oct/09/96
Lay Out & Set Batterboards	1.00 d	0.00 d	100	Sep/24/96	Sep/24/96
Excavate Turndown Edge	0.50 d	0.00 d	100	Sep/25/96	Sep/25/96
Set Edge Forms	1.50 d	0.00 d	100	Sep/25/96	Sep/26/96
HVAC Rough-in	0.50 d	0.00 d	100	Sep/25/96	Sep/25/96
Radon System Rough-in	0.50 d	0.00 d	100	Sep/25/96	Sep/25/96
Spread Gravel	0.50 d	0.50 d	0	Sep/30/96	Sep/30/96
Plumbing Rough-in	3.00 d	0.00 d	100	Sep/25/96	Sep/27/96
Termite Treatment	0.50 d	0.50 d	0	Sep/30/96	Sep/30/96
Set Grade Stakes	0.50 d	0.50 d	0	Oct/01/96	Oct/01/96
Place Poly and WWM	0.50 d	0.50 d	0	Oct/01/96	Oct/01/96
Inspect Slab	0.50 d	0.50 d	0	Oct/01/96	Oct/01/96
Place & Finish Concrete	1.00 d	1.00 d	0	Oct/02/96	Oct/02/96
Place Garage/Storage Slab	3.00 d	3.00 d	0	Oct/03/96	Oct/07/96
Place Concrete Patio	2.00 d	2.00 d	0	Oct/08/96	Oct/09/96
Structure Phase	26.50 d	20.66 d	22	Sep/23/96	Oct/30/96
Structure Dry-in Phase	27.50 d	16.50 d	40	Sep/23/96	Oct/31/96
Finish Phase	54.50 d	53.95 d	1	Sep/20/96	Dec/11/96
Finish Sitework Phase	7.50 d	7.50 d	0	Nov/08/96	Nov/20/96
Project Close-out Phase	2.00 d	2.00 d	0	Dec/11/96	Dec/13/96

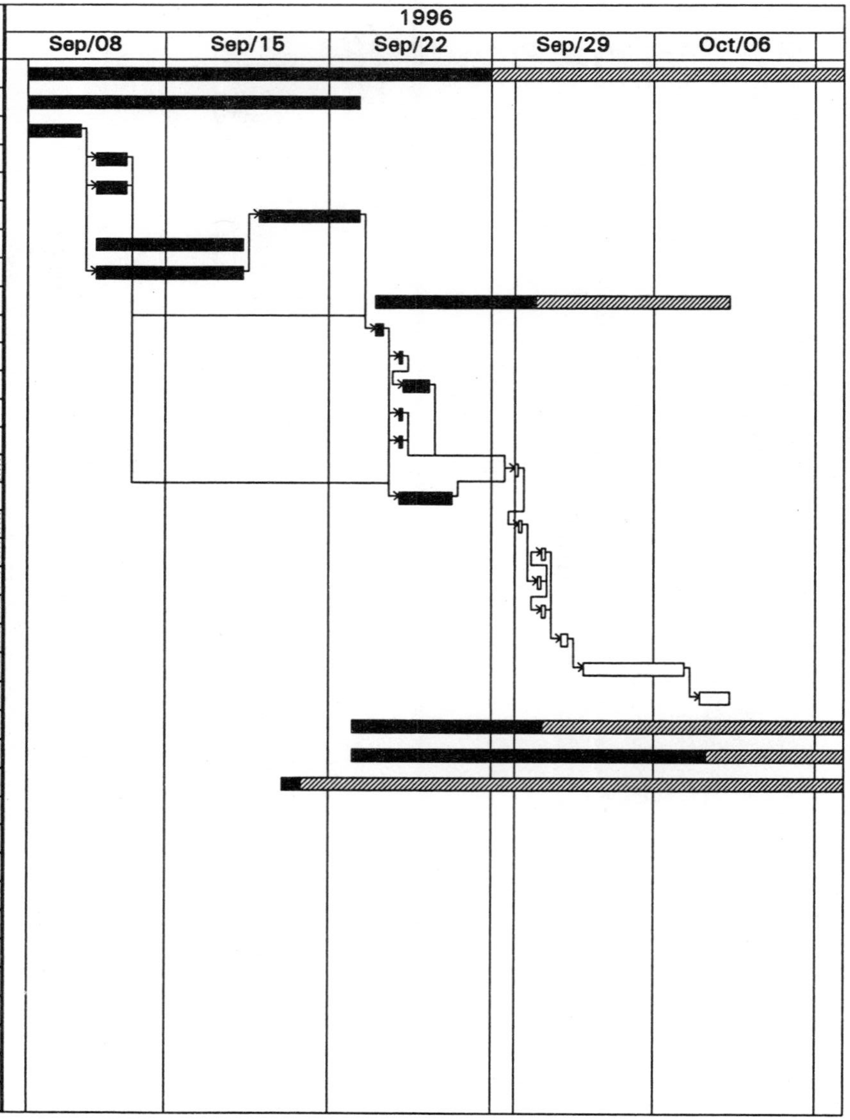

1996: Sep/08 | Sep/15 | Sep/22 | Sep/29 | Oct/06

Shaded activities are on the critical path.

First, the schedule can be updated to identify what effect the delays will have on the finish date of the project. The simplest way to do this is to increase the durations of the affected activities to reflect the delays, then manually recalculate the project completion date. (When revising the original dates, be sure to identify these changes so that everyone knows what is happening.) This method assumes that the original schedule is based on a good plan. The decision is then up to the builder to communicate to everyone that a change in the project schedule is going to occur.

Such a change can cause tremendous problems for a builder who has not created a working team and communicated with subcontractors. Everyone expects some change and most subs can adjust their schedules if given fair warning. Knowing that a change is coming helps people manage; not knowing can create problems.

The second method is to continue to work with the original schedule and push everyone to make up the lost time. This method can result in overtime expenses and bad relationships between the parties. The team should be trying to complete the project according to the time allowed; however, using the schedule as a stick against subcontractors is ineffective in solving the actual problem.

The builder should use the schedule as a tool to identify the party responsible for the activity and have them make adjustments to solve the scheduling problem. This approach requires commitment from the builder and the subcontractors to make the schedule work. If anyone on the team does not agree, it will be difficult to use the schedule as a positive tool.

Weather-Related Changes

Weather will always affect project completion. Projects have been built during seasons so dry that water had to be trucked to the site so the soil could be compacted. Other projects have had so much rain that materials have disappeared into the mud onsite. Builders and remodelers have to monitor weather predictions closely during every project.

Builders and remodelers can approach weather-related schedule complications by incorporating buffers into the original schedule. This can be done in two ways. The first method is to assign slightly longer durations to activities that are affected by bad weather. Ideally, the site grading activity in our example should take three days to complete. The five-day duration we have assigned allows for some delays caused by weather problems.

If the weather is good, the buffer allows the work to be completed a little ahead of schedule. If other tasks can be moved up, the time saved can be used to accelerate or provide a buffer for the rest of the project. Owners usually do not complain about a task or a project finishing early; however, to be on the safe side, it is best to stay with the original completion date. That way you are covered if a later delay uses up the time saved.

Subcontractors and suppliers may have problems accelerating their work if the project is progressing ahead of schedule. Their planning is set and changing it can affect other work they have in progress. Suppliers may have the material on order and be unable to change the delivery without added costs or problems. The best solution is to have a sound, realistic original schedule and stay on schedule.

The second method is to add a single activity at the end of the project to account for weather-related delays. This procedure has the same effect as doing nothing, except it provides the builder with a buffer when communicating with the owner. The duration of this activity should allow enough time for all weather-related delays. As these delays occur, the activity is credited with some percentage of completion to keep the project on schedule. We hope that at the end of the project some duration is remaining on this activity.

Some builders do not make schedule adjustments for weather delays. They feel that this creates a feeling of urgency among the people working on the project. However, because the schedule soon shows the project falling behind, the builder runs the risk that workers and subcontractors will simply stop using the schedule. The schedule will ultimately have to be adjusted to account for the delays. Of course, if the schedule is updated regularly, the updates will reflect past adverse weather conditions and other delays.

Both methods described above for handling weather have advantages and disadvantages. The first method is preferred because it provides a schedule that should be realistic and workable. The second method causes the builder to work harder and gives dates that will most likely need to be adjusted at some point. The bottom line is that if you create a schedule without some allowance for weather-related delays, the odds of finishing according to the original schedule are low.

Control

Control is defined as exercising restraint or directing influence over a subject. With regard to scheduling, the concept of controlling a project can also include the ability to check, test, or verify by parallel experiences. We exercise control to the extent that we can verify what is happening on the project compared with what is on the schedule. The updated schedule shows us whether activities at the jobsite are progressing according to our plan. When activities lag behind, the schedule helps us identify solutions that will cause the least disruption.

On our house schedule, we have set up order and delivery activities for most materials to closely precede their installation. Hopefully, the materials will be on the project when needed and will not need to be stored for any length of time. The initial order and delivery dates should be checked when updating the schedule.

An easy way to accomplish this is to have a small chart of just the order and delivery activities. Manually, you would have to separate these activities onto a new chart and update the chart as the changes occur. Using the *sort* feature available in most computer programs, you can easily transfer these activities onto a separate list.

Another way to separate the material order and delivery activities is to create a construction phase that groups all these activities together so the builder can easily review their status. Figure 4.5 shows an order and deliver phase.

A specific report like Figure 4.5 gives the builder important information concerning the current status of material deliveries. The start and finish times for installation activities reflect the *latest* time that the material might arrive onsite. Remember, materials that are ordered early can be scheduled for delivery as needed. The phase report only needs to contain items that might, if delayed, affect the finish date of the project.

Figure 4.5 can be adapted to include a notation of the responsible parties, so the builder has a quick reference regarding whom to ask about the status of each task. The materials delivery activities also can be sorted by subcontractor and the reports given to the appropriate subcontractors. If the schedule includes information about assigned costs, the builder or remodeler can use the sort function to generate internal reports that help monitor cash payments to subcontractors. Figure 4.6 shows the electrician's activities with the associated costs assigned.

The electrician has five activities to complete, three of which have assigned costs. (The total cost equals the subcontractor's contract amount as reflected in the project estimate.)

The schedule can assist in control by triggering the payments made to the subcontractor. The builder controls the money, the schedule, and the subcontractors. Naturally, the sub-

Figure 4.5: Order and Deliver Phase

Phase/Activity List	Resp	Duration	Start	End
Order/Deliver Phase		53.50 d	Sep/23/96	Dec/11/96
Order Framing Material	Builder	2.00 d	Sep/27/96	Sep/30/96
Deliver Framing Material	Builder	1.00 d	Oct/01/96	Oct/01/96
Select Whirlpool Tub	Owner	1.00 d	Sep/23/96	Sep/23/96
Order & Deliver Tubs	Builder	4.00 d	Sep/24/96	Sep/27/96
Order Ext Doors & Win	Builder	10.00 d	Sep/23/96	Oct/04/96
Deliver Ext Doors & Win	Builder	1.00 d	Oct/07/96	Oct/07/96
Select Brick	Owner	1.00 d	Sep/24/96	Sep/24/96
Order Brick	Builder	1.00 d	Sep/25/96	Sep/25/96
Deliver Brick	Builder	15.00 d	Sep/26/96	Oct/17/96
Deliver Synthetic Stucco	Stucco Sub	0.50 d	Oct/15/96	Oct/15/96
Order Sheetrock	Builder	1.00 d	Oct/17/96	Oct/17/96
Deliver Sheetrock	Builder	1.00 d	Oct/18/96	Oct/18/96
Order & Deliver Int Trim	Builder	2.00 d	Oct/21/96	Oct/22/96
Select Entry Marble	Owner	1.00 d	Sep/24/96	Sep/24/96
Order Marble	Builder	1.00 d	Sep/25/96	Sep/25/96
Deliver Marble	Floor Sub	40.00 d	Sep/26/96	Nov/22/96
Order Appliances	Builder	1.00 d	Oct/15/96	Oct/15/96
Deliver & Set Appliances	Builder	1.00 d	Dec/10/96	Dec/11/96
Order Cabinets	Builder	1.00 d	Oct/11/96	Oct/11/96
Build & Deliver Cabinets	Cab Sub	13.00 d	Oct/15/96	Oct/31/96
Select Wallpaper	Owner	1.00 d	Oct/23/96	Oct/23/96
Order Wallpaper	Builder	1.00 d	Oct/24/96	Oct/24/96
Deliver Wallpaper	Painter	9.00 d	Oct/25/96	Nov/06/96

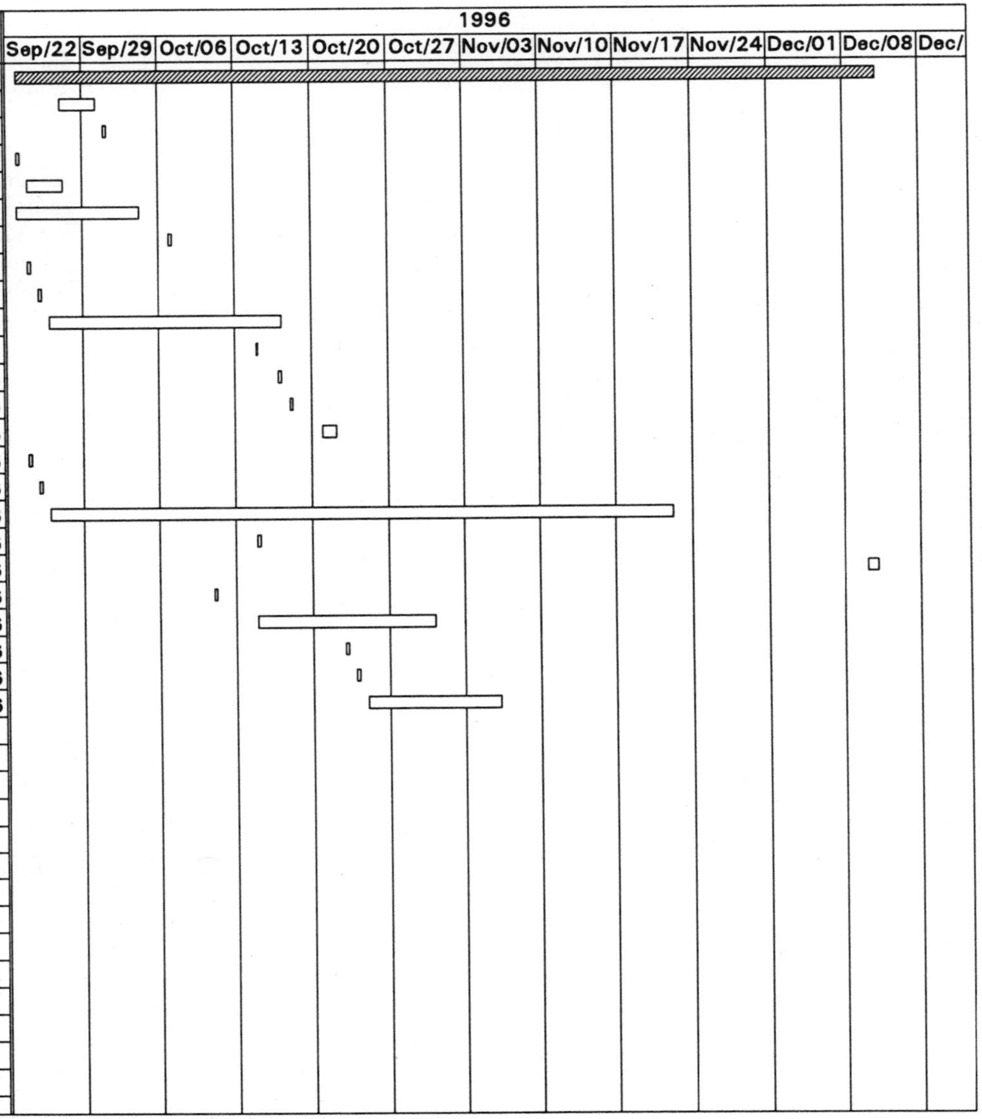

1996

Sep/22 Sep/29 Oct/06 Oct/13 Oct/20 Oct/27 Nov/03 Nov/10 Nov/17 Nov/24 Dec/01 Dec/08 Dec/

Figure 4.6: Electrical Subcontractor's Activities with Costs Assigned

Phase/Activity List	Cost	Resp	Duration	Start	End
The House 1 Schedule	126,788.00		65.50 d	Sep/09/96	Dec/13/96
Project Startup Phase	$825.00		11.00 d	Sep/09/96	Sep/23/96
Obtain Temporary Power	$50.00	Electrician	2.00 d	Sep/12/96	Sep/13/96
Structure Phase	$30,660.00		26.50 d	Sep/23/96	Oct/30/96
Electrical Rough-in	$4,181.00	Electrician	2.00 d	Oct/23/96	Oct/24/96
Electrical Inspection	$0.00	Electrician	0.50 d	Oct/25/96	Oct/25/96
Finish Phase	$48,472.00		52.50 d	Sep/24/96	Dec/11/96
Electrical Trim-out	$4,137.00	Electrician	1.00 d	Dec/06/96	Dec/09/96
Final Electrical Inspection	$0.00	Electrician	0.50 d	Dec/09/96	Dec/09/96

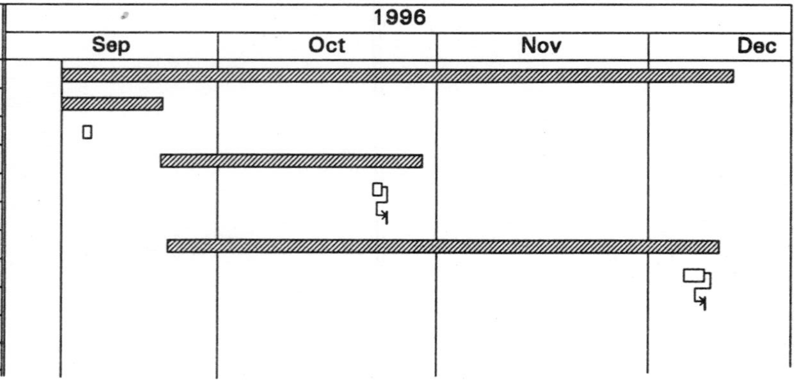

Figure 4.7: Framing Subcontractor's Activities

Project/Phase/Activity List	Resp	Duration	Start	End
The House 1 Schedule		65.50 d	Sep/09/96	Dec/13/96
Structure Phase		26.50 d	Sep/23/96	Oct/30/96
Wall and Sheathing	Framing	3.00 d	Oct/03/96	Oct/07/96
Roof Framing & Decking	Framing	5.00 d	Oct/10/96	Oct/17/96
Install Deadwood and Blk	Framing	0.50 d	Oct/18/96	Oct/18/96
Structure Dry-in Phase		27.50 d	Sep/23/96	Oct/31/96
Install Fascia and Soffits	Framing	4.00 d	Oct/18/96	Oct/24/96
Install Ext Doors & Win	Framing	4.00 d	Oct/18/96	Oct/24/96
Install Overhead Door	Framing	0.50 d	Oct/24/96	Oct/24/96
Install Siding	Framing	0.50 d	Oct/25/96	Oct/25/96

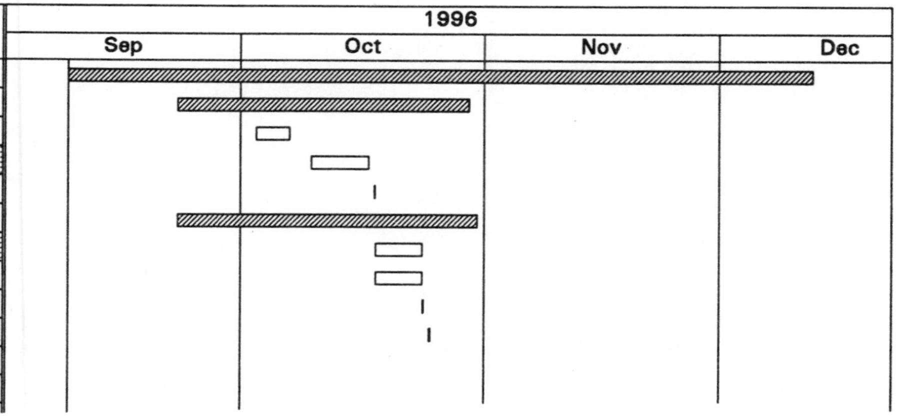

Shaded activities are on the critical path.

contractors will pay attention to the tasks that are tied to payments. If these tasks are not represented in the schedule, the subcontractors may view the schedule as a lower priority. If payments are tied to the schedule, completion of activity triggers authorization of the corresponding payment. If the activity is *not* completed, the payment can be delayed until completion (provided the builder and subcontractor agreed to the activities and the payments at the outset of the job).

Many builders and remodelers do not pay subcontractors by a cost-loaded schedule. If payments are specified for activities in a contract, the builder must use the same activities in the schedule for monitoring progress and invoicing. The motives and rationale behind payment schedules may be different from those behind work planning and scheduling. Also, if the project involves many change orders, keeping the costs and the schedule coordinated can become cumbersome.

This type of cost management is not for everyone and cost-loading should be added to the planning and scheduling process only after the builder or remodeler has become comfortable with the basic scheduling process. Cost management based on poor scheduling diagrams can cause problems and should be avoided. If you typically have few changes in your jobs, and have a scheduling system that is working well, however, cost-loaded schedules may work well for you. (For more on cost-loading see Cash Flow later in this chapter.)

You can also use the schedule to assist with resource management. This is the process of identifying needs for—and the availability of—special or limited resources. Resources can be personnel, equipment, or materials. For example, steel formwork, a backhoe, and a finish carpenter are all resources. There are two ways that the schedule can be used to manage resources.

The first technique is to build soft restraints into the schedule that force activities to occur in an orderly linear sequence. Resources can then be assigned to activities with the knowledge that a person or piece of equipment will not be in two places at the same time. Unfortunately, with only one activity in progress at any given time, the building process is often inefficient and wastes valuable time.

The second technique is to plan the work in detail and then review the schedule to detect conflicts in resources. This is done by identifying activities that require the same resource and checking to see if the activities overlap on the schedule. Some of the better computer scheduling programs will automatically identify overlapped or *overloaded* resources. Figure 4.7 shows the activities of our framing subcontractor.

Most overloads come from using the same subcontractor on several projects at the same time. The subcontractor with one crew is then expected to be in two places at the same time. Resource constraints caused by commitments to multiple projects can be a big scheduling problem, and will be discussed later in this chapter.

Figure 4.7 shows that the framing work is scheduled to begin with the walls and sheathing, continue with the fascia and soffits, and end with the siding being installed. The framing crew stays busy during the month of October and should be finished by October 25. However, the framing crew is required to be working on the fascia and soffits and installing the doors and windows at the same time. With only one crew, this is a resource overload that will prevent the framing crew from staying on schedule.

The builder can explore at least three options to correct this overload. The first is to add a restraint that causes one activity to be completed before the other activity. Since the *critical* activity is the fascia and soffits, this should go first. To look at how this strategy works, the builder would place the restraint, then recalculate the forward and backward pass, and determine the effect on the schedule.

If one activity has a shorter duration than the other, it may be more effective to complete the shorter activity first. This solution frees the resources with smaller impact; however, it can cause problems if the new sequence of activities delays other subs' critical activities. Checking this solution by redoing the network calculations will reveal whether the approach creates any major problems.

A third option is to discuss the problem with the subcontractor and determine whether any additional resources may be available to complete the work according to the original schedule. Depending on the situation, adding resources may cost more money; however, it is sometimes the best approach, particularly if it prevents more costly problems down the line. When overloading occurs, it is important to discuss the problems and find the best solution as quickly as possible. If no action is taken, delays will be sure to occur, and the emergency solution is likely to cost much more than a solution worked out in the planning stage.

Cash Flow

Good cash flow is essential to survival in the construction business. The construction business has two parts—doing the work and getting paid. The money management of a project and a company can be enhanced by monitoring cash flow according to the schedule.

Most jobs are financed in one of three ways:

- Out of the builder's or remodeler's own funds
- By periodic payments from the remodeling customer or the owner of a presold house
- By funds borrowed from a bank or other lender

Payments from the owner or lender may be scheduled to occur as needed, weekly, monthly, or upon completion of specific project phases such as slab, dry-in, and trim-out. Banks may also schedule payments based on their own list of activities. Payments for extra items and changes may be added to the schedule.

The ability to schedule payments based on an accurate forecast of cash needs greatly helps builders or remodelers who are financing projects with their own funds. Remodeling clients and owners of presold houses also appreciate receiving a projected cash requirement schedule.

Likewise, banks are favorably impressed with builders and remodelers who have organized and projected their cash-flow needs. Of course, builders who supply an owner or banker with a projected cash requirement schedule are under pressure to produce. Therefore, basing the cash-flow projection on a sound schedule is extremely important.

The operation of dividing the project's direct *costs* among the scheduled activities was discussed in Chapter 3. These costs, however, are only part of the cash flow. Amounts for job overhead, general or home office overhead, and profit that are not directly assigned to activities also can be monitored in relation to the schedule. Many builders load overhead and profit arbitrarily at the front of the schedule to minimize their risk. If a more detailed distribution of the project's direct and indirect costs (including markup) would be helpful, three methods can be used to assign values to specific activities.

Earned Value Based on Cost. In the first method, an appropriate amount of money can be distributed to each activity based on some factor such as estimated cost, duration, or risk. Estimated cost and duration are fairly easy to project using the project estimate. For example, the builder can determine the estimated direct cost of the project without the markup (general overhead and profit) to be $100,000. If the total contract equals $150,000, a mark-up factor could be calculated by: Total Contract ÷ Cost = 1.5. The cost of each activity can then be multiplied by 1.5 to determine its earned value. (Remember, earned value represents the total value of the activity in relation to the project.) In our example, the earned value for the wall framing would be calculated as follows: The Frame Walls activity costs $5,000; therefore, the earned value of this activity equals $5,000 multiplied by 1.5, or $7,500.

Distribution by risk is another matter. Each activity can be assigned a dollar amount based on the *loss* that might be incurred were something to go wrong while trying to complete the activity. Critical activities, whose potential losses are higher, are assigned larger amounts of money than noncritical activities. Basically, you place the money where your risk is greatest.

Earned Value Based on Percentage. Earned value also can be factored into the schedule based on the percentage of work completed during the month. With this method, the amount added to the direct costs to determine the payment clearly shows the owner what money is considered overhead and profit. Many builders do not want to provide this information to owners or other parties. However, if you are building with a cost-plus contract or a fixed-fee management agreement, this method may be appropriate.

Earned Value Based on Even Distribution. A third method is to evenly distribute the costs, including the amount for overhead and profit, over the duration of the project. Each workday is allotted the same amount of money. This method is easy to calculate and places money in the schedule, but the results bear no relation to the completion of activities. While this method is better than nothing, it leaves the builder exposed to more risk than the other two methods.

The first method takes the most time to calculate, but allows the builder the most control over the payment schedule and cash flow. The second and third methods require less effort and, while they allow for less control of risks, they also allow the builder to keep overhead and profit separated from the actual costs of the activities. For schedules used internally only, these methods can help the builder to easily monitor the anticipated costs in relation to the anticipated income.

Preparing a Cost-to-Date Summary

For simplicity, we have used the third method of distribution in calculating a cost-to-date summary for our sample house. Figure 4.8 shows a summary of the activities and the costs as distributed across each phase. The bottom of the figure also displays the anticipated monthly amount to be billed to the owner. The $170,750 amount given for the total house schedule includes the total of all phases plus $43,962, an amount added to cover costs for the land, septic system, and closing costs. This allows us to include all project costs in the schedule for billing purposes.

The phase costs shown equal the direct costs of all the activities under that phase. The monthly totals for each activity are not shown, but they also have been broken out by the computer and calculated to provide the billing amounts at the bottom of the chart. The builder, the owner, and the banker can see how much money the builder plans to request at the end of each month.

At this point, we have only been dealing with the income side of the cash-flow picture. The expense side is just as important and can also be projected with the schedule. Figure 4.9 contains assigned costs for all activities. Removing the $83,750 from the "total house" line adjusts the schedule to represent only the expenses anticipated during the project.

Figure 4.8: Anticipated Income

Project/Phase List	Billings	Duration	Start	End
The House 1 Schedule	$170,750.00	65.50 d	Sep/09/96	Dec/13/96
Project Startup Phase	$825.00	11.00 d	Sep/09/96	Sep/23/96
Sitework Phase	$2,700.00	5.00 d	Sep/12/96	Sep/18/96
Slab-on-Grade Phase	$14,086.00	12.00 d	Sep/24/96	Oct/09/96
Structure Phase	$30,660.00	26.50 d	Sep/23/96	Oct/30/96
Structure Dry-in Phase	$21,145.00	27.50 d	Sep/23/96	Oct/31/96
Finish Phase	$48,472.00	52.50 d	Sep/24/96	Dec/11/96
Finish Sitework Phase	$8,600.00	7.50 d	Nov/08/96	Nov/20/96
Project Close-out Phase	$300.00	2.00 d	Dec/11/96	Dec/13/96

Period Billings

	Sep	Oct	Nov	Dec
1996	$24,289.21	$80,945.09	$34,717.29	$30,798.41

Scale: $0.00 – $100,000.00

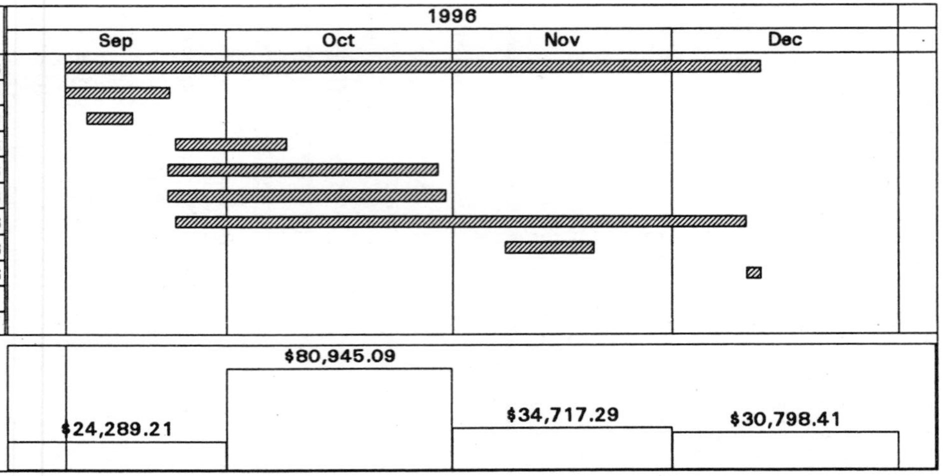

Note: The House Schedule total includes the sum of all the phases plus an additional amount to cover the cost of the land, septic system, and closing costs. The period billings equal the total costs broken out by month based on completion of activities.

Figure 4.9: Anticipated Expenses

Project/Phase List	Costs	Duration	Start	End
The House 1 Schedule	$129,217.00	65.50 d	Sep/09/96	Dec/13/96
Project Startup Phase	$825.00	11.00 d	Sep/09/96	Sep/23/96
Sitework Phase	$2,700.00	5.00 d	Sep/12/96	Sep/18/96
Slab-on-Grade Phase	$14,086.00	12.00 d	Sep/24/96	Oct/09/96
Structure Phase	$30,660.00	26.50 d	Sep/23/96	Oct/30/96
Structure Dry-in Phase	$21,145.00	27.50 d	Sep/23/96	Oct/31/96
Finish Phase	$48,472.00	52.50 d	Sep/24/96	Dec/11/96
Finish Sitework Phase	$8,600.00	7.50 d	Nov/08/96	Nov/20/96
Project Close-out Phase	$500.00	2.00 d	Dec/11/96	Dec/13/96

Period Costs

	Sep	Oct	Nov	Dec
1996	$14,641.65	$67,350.80	$21,561.53	$25,663.02

Scale: $0.00 – $84,000.00

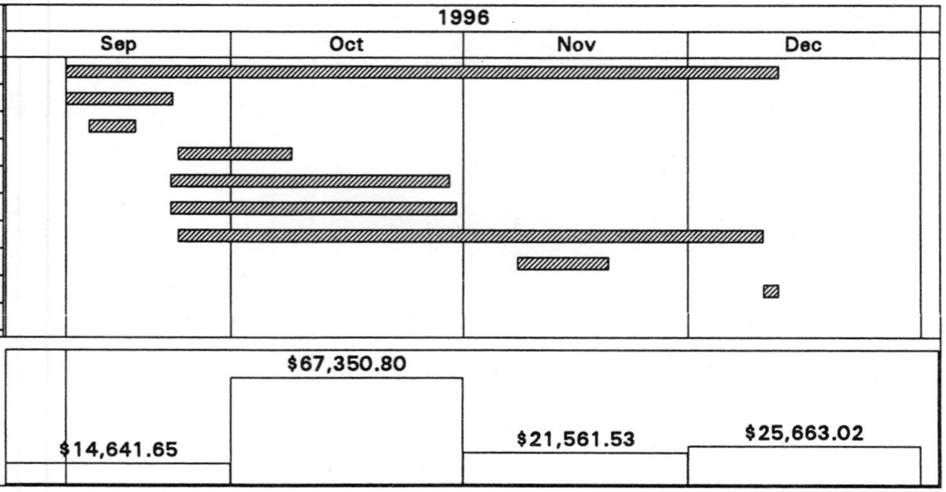

Note: The $129,217 cost reflects the base-bid cost of $254,500 less the land, closing costs, and costs for Builders Risk Insurance, overhead, and profit.

Of course, not all bills are paid on the same schedule. Some, such as payroll and draws from subcontractors, are usually paid weekly. Other bills, such as material invoices, are paid monthly. To keep our example simple, we will analyze the expenses on a monthly basis.

Preparing a Cash-Flow Forecast

Figure 4.10 shows the balance of the projected income and expenses on a monthly basis. (Note: numbers have been rounded.) If we assume that the builder will receive monthly payments from the owner based on the work completed, the cash flow should appear similar to the information in Figure 4.10.

Figure 4.10: Payment Schedule

Month	September	October	November	December
Expense	$14,642	$67,351	$21,562	$25,662
Income	$24,289	$80,945	$34,717	$30,799
Balance	$9,647	$13,594	$13,155	$15,137

The builder is working with a positive cash position each month during the life of this project. However, weekly payrolls and subcontractor bills may require the builder to spend money before receiving income from the monthly billing. To avoid possible cash-flow problems, the builder may contract with the owner to make bimonthly payments. Alternatively, the builder may contract with the subs to pay them on a monthly rather than a weekly basis. To the extent the builder can match income with expenses across the schedule, he or she will have reduced the risk of cash-flow problems.

The cash-flow picture presented in Figure 4.10 is only an estimate, and it is only as good as the schedule. If project delays occur, the wise builder will redo this chart. Delays tend to reduce the short-term cash flow and increase the cost of completing the project, which may cause problems at completion.

An initial diagram gives an idea as to the expectations of what is ahead. As the project progresses, a new diagram can be constructed to see how much change has occurred. The diagrams are as accurate as the information put into the schedule.

EXPEDITING (CRASHING) A SCHEDULE

When a project falls behind schedule, builders and remodelers may take action to expedite or "crash" the schedule. Basically, this is a process of systematically reducing the durations of remaining activities or revising the order of activities to complete the project on time.

If the builder cannot change the durations or revise the sequence and have the project finish on schedule, then increasing the number of workers, working overtime, shortening delivery times, and improving coordination between subcontractors (and more closely monitoring their activities) may salvage the schedule. However, these actions usually result in increased costs.

The increased cost associated with crashing a schedule is difficult to determine and should be carefully calculated. Subcontractors and suppliers have committed to work according to a plan and the plan has changed. How much will it cost to change?

In our house example, let's say that framing lumber has not been delivered. Our supplier has informed us that the lumber will arrive three weeks later than the planned date of October 1. Framing material will now be scheduled for delivery on October 21. The effect of this delay is shown in Figure 4.11.

The wide, dark gray bars in Figure 4.11 represent the times the activities were *originally* scheduled. The clear bars represent the new dates the activities will take place. The overall effect of the delay, as shown by the lighter gray bars on the summary activity, is a one-week delay in completion. The project close-out phase has a new end date of December 30, which means the project will now finish after Christmas—and the owner is not happy with this situation.

In Figure 4.11, the effects of the delay have been recalculated to adjust the locations of all activities in the diagram. To make this adjustment manually, the builder would have to recalculate the forward and backward pass and redraw the diagram.

The most effective way to approach crashing the schedule is to think strategically. Look at which activities will make the most difference in recapturing the lost time. First, is the material available from another supplier? The material cost may be greater, but accepting the delay, disruption, and rescheduling may have a greater cost.

Second, consider the activities being done with your own forces. Finishing the slab earlier will not do any good, as the slab would then sit there for two weeks while you wait for the framing lumber to arrive. It is illogical to spend additional money on activities that will not help correct the problem.

On the other hand, Wall and Sheathing and Roof and Deck are prime activities for consideration. Adding more workers to these activities will reduce their duration. We will double the crews and—we hope—reduce the duration in half. However, there is a practical limit to adding more workers. At some point, the workers begin to get in each others way or they cannot be managed efficiently.

Besides adding workers, we can overlap activities. The roof framing can begin after a portion of the wall has been erected. This overlapping will save time, but only with close coordination and extra supervision. It is also important to avoid unsafe working conditions. For example, avoid having workers doing overhead work that places others at risk.

We have halved the durations for wall and sheathing and roof and deck installation, and have decided to overlap the roof framing with the installation of the walls. The modified schedule in Figure 4.12 shows that these revisions have brought the overall schedule back to a completion date of December 20. At this point, we would decide whether more changes need to be made or if the schedule is acceptable. We will choose to work with the schedule as shown, noting that many of the dry-in phase and finish phase activities still have to shift to accommodate the changes we have made, and the subcontractors will need to be notified.

The next step in crashing the schedule is to fine tune activities to reduce the number of workers and subcontractors affected by the change. In our example, the soffit and exterior windows and doors will still take four days each to finish. Since the additional people are already onsite, will it help to crash these activities? Is this important? If so, we could increase the crew sizes; if not, we could reschedule the mason and stucco subcontractors. Either way, the change will not affect the overall duration of the project.

Figure 4.11: Framing Material Delay

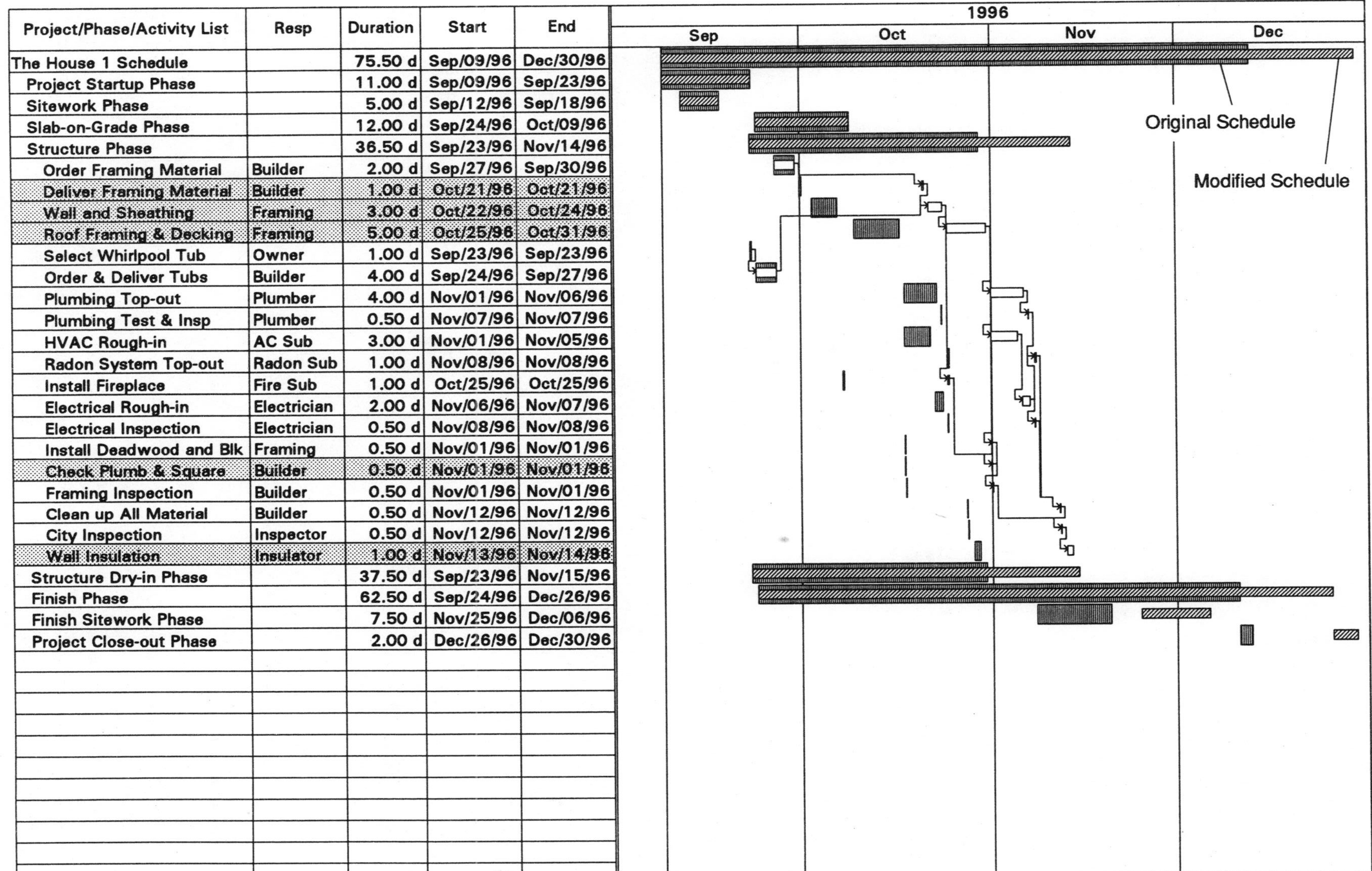

Project/Phase/Activity List	Resp	Duration	Start	End
The House 1 Schedule		75.50 d	Sep/09/96	Dec/30/96
Project Startup Phase		11.00 d	Sep/09/96	Sep/23/96
Sitework Phase		5.00 d	Sep/12/96	Sep/18/96
Slab-on-Grade Phase		12.00 d	Sep/24/96	Oct/09/96
Structure Phase		36.50 d	Sep/23/96	Nov/14/96
Order Framing Material	Builder	2.00 d	Sep/27/96	Sep/30/96
Deliver Framing Material	Builder	1.00 d	Oct/21/96	Oct/21/96
Wall and Sheathing	Framing	3.00 d	Oct/22/96	Oct/24/96
Roof Framing & Decking	Framing	5.00 d	Oct/25/96	Oct/31/96
Select Whirlpool Tub	Owner	1.00 d	Sep/23/96	Sep/23/96
Order & Deliver Tubs	Builder	4.00 d	Sep/24/96	Sep/27/96
Plumbing Top-out	Plumber	4.00 d	Nov/01/96	Nov/06/96
Plumbing Test & Insp	Plumber	0.50 d	Nov/07/96	Nov/07/96
HVAC Rough-in	AC Sub	3.00 d	Nov/01/96	Nov/05/96
Radon System Top-out	Radon Sub	1.00 d	Nov/08/96	Nov/08/96
Install Fireplace	Fire Sub	1.00 d	Oct/25/96	Oct/25/96
Electrical Rough-in	Electrician	2.00 d	Nov/06/96	Nov/07/96
Electrical Inspection	Electrician	0.50 d	Nov/08/96	Nov/08/96
Install Deadwood and Blk	Framing	0.50 d	Nov/01/96	Nov/01/96
Check Plumb & Square	Builder	0.50 d	Nov/01/96	Nov/01/96
Framing Inspection	Builder	0.50 d	Nov/01/96	Nov/01/96
Clean up All Material	Builder	0.50 d	Nov/12/96	Nov/12/96
City Inspection	Inspector	0.50 d	Nov/12/96	Nov/12/96
Wall Insulation	Insulator	1.00 d	Nov/13/96	Nov/14/96
Structure Dry-in Phase		37.50 d	Sep/23/96	Nov/15/96
Finish Phase		62.50 d	Sep/24/96	Dec/26/96
Finish Sitework Phase		7.50 d	Nov/25/96	Dec/06/96
Project Close-out Phase		2.00 d	Dec/26/96	Dec/30/96

The wide darker grey bars show the original schedule. The lighter grey bars show how the delay in delivery of framing materials affects all subsequent activities on the critical path. Start and end dates have been adjusted to reflect the delay. Shaded activities in the Activity List are on the critical path.

Figure 4.12: Adjusted Framing Activities

Project/Phase/Activity List	Resp	Duration	Start	End
The House 1 Schedule		70.50 d	Sep/09/96	Dec/20/96
Project Startup Phase		11.00 d	Sep/09/96	Sep/23/96
Sitework Phase		5.00 d	Sep/12/96	Sep/18/96
Slab-on-Grade Phase		12.00 d	Sep/24/96	Oct/09/96
Structure Phase		31.50 d	Sep/23/96	Nov/06/96
Order Framing Material	Builder	2.00 d	Sep/27/96	Sep/30/96
Deliver Framing Material	Builder	1.00 d	Oct/21/96	Oct/21/96
Wall and Sheathing	Framing	1.50 d	Oct/22/96	Oct/23/96
Roof Framing & Decking	Framing	2.50 d	Oct/22/96	Oct/24/96
Select Whirlpool Tub	Owner	1.00 d	Sep/23/96	Sep/23/96
Order & Deliver Tubs	Builder	4.00 d	Sep/24/96	Sep/27/96
Plumbing Top-out	Plumber	4.00 d	Oct/25/96	Oct/30/96
Plumbing Test & Insp	Plumber	0.50 d	Oct/31/96	Oct/31/96
HVAC Rough-in	AC Sub	3.00 d	Oct/25/96	Oct/29/96
Radon System Top-out	Radon Sub	1.00 d	Nov/01/96	Nov/01/96
Install Fireplace	Fire Sub	1.00 d	Oct/23/96	Oct/24/96
Electrical Rough-in	Electrician	2.00 d	Oct/30/96	Oct/31/96
Electrical Inspection	Electrician	0.50 d	Nov/01/96	Nov/01/96
Install Deadwood and Blk	Framing	0.50 d	Oct/25/96	Oct/25/96
Check Plumb & Square	Builder	0.50 d	Oct/25/96	Oct/25/96
Framing Inspection	Builder	0.50 d	Oct/25/96	Oct/25/96
Clean up All Material	Builder	0.50 d	Nov/04/96	Nov/04/96
City Inspection	Inspector	0.50 d	Nov/04/96	Nov/04/96
Wall Insulation	Insulator	1.00 d	Nov/05/96	Nov/06/96
Structure Dry-in Phase		32.50 d	Sep/23/96	Nov/07/96
Finish Phase		57.50 d	Sep/24/96	Dec/18/96
Finish Sitework Phase		7.50 d	Nov/18/96	Nov/27/96
Project Close-out Phase		2.00 d	Dec/18/96	Dec/20/96

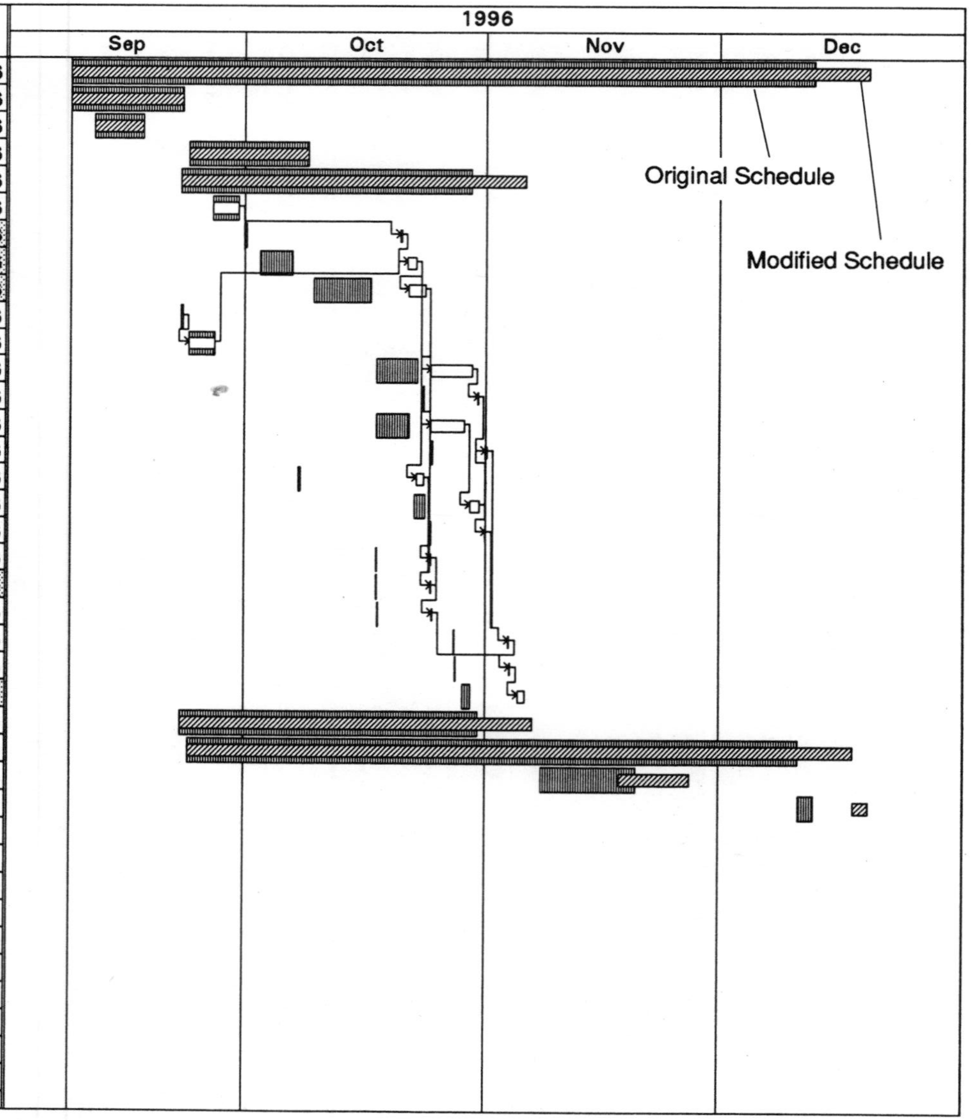

Shaded activities are on the critical path.

A crashed schedule requires subcontractors to adjust their own project schedules. You may be asking them to commit additional time and workers to your project or to work under conditions that differ from those on which they based their original estimates. Depending on the subcontractor and on the degree of change, you may have to pay some subcontractors to accelerate their work.

The builder or remodeler and the subs must work together as a team if the crashed schedule is to succeed. When changes and delays affect the project, it is the team that will make corrections and finish on time. Everyone involved has to make adjustments. If the adjustments result in added costs, the relationships between the parties will determine who will pay the added costs and how much.

A builder may have to pay subcontractors for changes in the schedule, especially if the builder often has to ask subs to revise or accelerate their work plans. Builders who plan well seldom ask their subcontractors to accelerate and, when they do, their subcontractors often will cooperate at no additional cost if possible. Some builders with tight cost constraints will let the schedule slip rather than pay extra. Good working relationships, good communication, and good scheduling will help reduce most of these problems.

MULTIPLE PROJECT SCHEDULING

To manage multiple projects, the first step is to develop a detailed schedule for each project. An overall master schedule can then be developed by combining the schedules. The overall schedule should be updated and managed with the same regularity as the individual schedules. If you are using a computerized scheduling program, the overall schedule will automatically reflect the updates made to the individual schedules. As discussed earlier, to be most effective, the updating should be done on a weekly basis.

To construct a multiproject schedule, we will assume that we are building three more houses, minor variations on our sample house, in the same area. House 2 will start on October 7, House 3 on October 21, and House 4 on November 4.

Initial schedules for each of the three additional houses can be made by copying the original house schedule and adjusting the start dates. The schedules can then be combined into a single master schedule as shown in Figure 4.13.

To create the master schedule manually, the builder can draw individual diagrams for each project (or draw them all on one large sheet of paper) and lay them out together to compare them. Because such diagrams can become extremely large, the builder may wish to only show the critical activities for each project.

Multiproject scheduling can be a nightmare. The slightest change can require rethinking the logic, recalculation, and redrawing large sections of the logic diagrams. Some builders find it an easier approach to send the individual project schedules to the subcontractors and suppliers and ask if they see any conflicts or problems. Since each subcontractor handles a small portion of the work, each can quickly review and respond with changes or suggestions. Depending on the complexity of the scheduling information the builder provides, he or she may have to show the subcontractor how to read the diagram. Also, even if the builder delegates some of the problem-solving tasks to the subs, the builder or remodeler must still review all the options and assess how best to allocate resources across the various jobs.

A master schedule allows builders and remodelers to monitor the way company and subcontractor resources have been assigned. For example in Figure 4.13, we can quickly see that the electrician's activities can be coordinated among the three projects. The sort function on the computer can be used to create a new chart focusing on only the electrician's activities (Figure 4.14). Manually, the same information could be highlighted, re-listed, or redrawn.

Figure 4.14 reveals that the electrician will have one conflict near the beginning of December. The electrical rough-in for House 3 and the electrical trim-out for House 1 are both scheduled to be worked on during the same time period.

Since the problem is still three months away, the best course of action is simply to alert the electrician to the potential conflict and wait. During the next three months, a lot of changes are likely to occur and the problem may simply correct itself. Additionally, the electrical subcontractor might have other projects and crews that may be free to do the work. Adjusting the schedule in an attempt to prevent this overlap at this point in time is a waste of effort. Consistent communication with the subcontractors, however, will keep everyone prepared to deal with this issue. A decision as to how the conflict will be handled should be made three to four weeks before the work is to be started.

Another important subcontractor whose work schedule should be checked is the framing subcontractor. This work represents a larger portion of the critical activities in most schedules and delays will cause problems for everyone. Figure 4.15 shows the activities we have assigned to the framing subcontractor.

The framing subcontractor is scheduled to work steadily from October 3 until December 24. The graph shows that several times during these projects, the framer will be on two projects. The fascia and soffit activity for House 2 is at the same time the wall framing for House 3 is scheduled. This same conflict occurs between House 3 and House 4.

We have several options for addressing this problem. First, activities can be rearranged to eliminate the overlap. Second, we can ignore the problem and hope that the framing subcontractor will have the crews to complete the work on time. A third response could be to alert the carpenter to the conflicts and discuss various strategies for completing the work, such as increasing the crews or working overtime during certain weeks. Lastly, we could hire another framing sub to provide additional help. There are also other strategies for avoiding the delays that the overlapping activities may cause. The main point is that identifying these scheduling conflicts allows the builder time to make adjustments *before* the conflict becomes a problem at the jobsite.

Combining all active projects into a master schedule gives the builder or remodeler a picture of the total work being completed. It can help show when new work must be started and it can help spot potential conflicts between resources before they occur.

Multiproject scheduling can be time-consuming and assistance may be required in planning, scheduling, and monitoring. Consultants are available with experience to assist. Many universities and the National Association of Home Builders have materials available that can help improve your command of the scheduling process.

Figure 4.13: Master Schedule for Four Projects

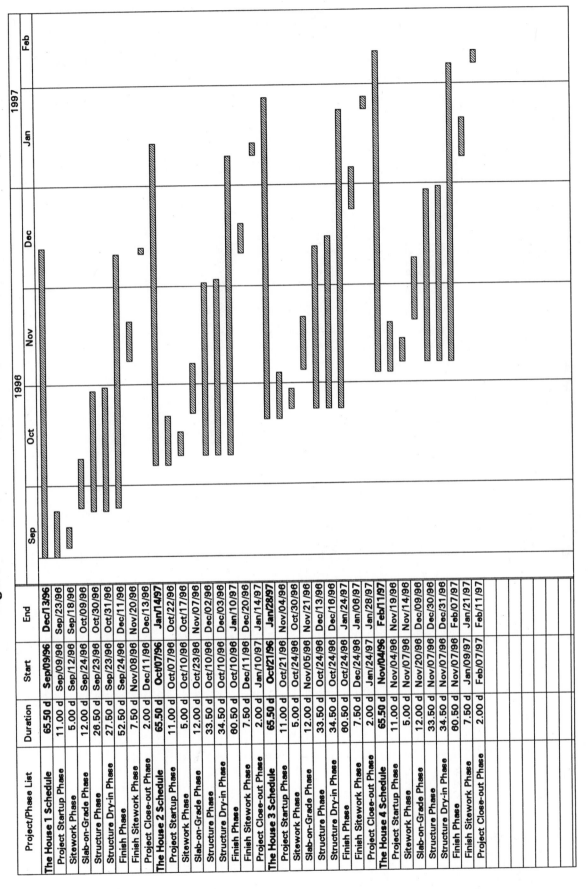

Project/Phase List	Duration	Start	End
The House 1 Schedule	**65.50 d**	**Sep/09/96**	**Dec/13/96**
Project Startup Phase	11.00 d	Sep/09/96	Sep/23/96
Sitework Phase	5.00 d	Sep/12/96	Sep/18/96
Slab-on-Grade Phase	12.00 d	Sep/24/96	Oct/09/96
Structure Phase	26.50 d	Sep/23/96	Oct/30/96
Structure Dry-in Phase	27.50 d	Sep/23/96	Oct/31/96
Finish Phase	52.50 d	Sep/24/96	Dec/11/96
Finish Sitework Phase	7.50 d	Nov/08/96	Nov/20/96
Project Close-out Phase	2.00 d	Dec/11/96	Dec/13/96
The House 2 Schedule	**65.50 d**	**Oct/07/96**	**Jan/14/97**
Project Startup Phase	11.00 d	Oct/07/96	Oct/22/96
Sitework Phase	5.00 d	Oct/10/96	Oct/17/96
Slab-on-Grade Phase	12.00 d	Oct/23/96	Nov/07/96
Structure Phase	33.50 d	Oct/10/96	Dec/02/96
Structure Dry-in Phase	34.50 d	Oct/10/96	Dec/03/96
Finish Phase	60.50 d	Oct/10/96	Jan/10/97
Finish Sitework Phase	7.50 d	Dec/11/96	Dec/20/96
Project Close-out Phase	2.00 d	Jan/10/97	Jan/14/97
The House 3 Schedule	**65.50 d**	**Oct/21/96**	**Jan/28/97**
Project Startup Phase	11.00 d	Oct/21/96	Nov/04/96
Sitework Phase	5.00 d	Oct/24/96	Oct/30/96
Slab-on-Grade Phase	12.00 d	Nov/05/96	Nov/21/96
Structure Phase	33.50 d	Oct/24/96	Dec/13/96
Structure Dry-in Phase	34.50 d	Oct/24/96	Dec/16/96
Finish Phase	60.50 d	Oct/24/96	Jan/24/97
Finish Sitework Phase	7.50 d	Dec/24/96	Jan/06/97
Project Close-out Phase	2.00 d	Jan/24/97	Jan/28/97
The House 4 Schedule	**65.50 d**	**Nov/04/96**	**Feb/11/97**
Project Startup Phase	11.00 d	Nov/04/96	Nov/19/96
Sitework Phase	5.00 d	Nov/07/96	Nov/14/96
Slab-on-Grade Phase	12.00 d	Nov/20/96	Dec/09/96
Structure Phase	33.50 d	Nov/07/96	Dec/30/96
Structure Dry-in Phase	34.50 d	Nov/07/96	Dec/31/96
Finish Phase	60.50 d	Nov/07/96	Feb/07/97
Finish Sitework Phase	7.50 d	Jan/09/97	Jan/21/97
Project Close-out Phase	2.00 d	Feb/07/97	Feb/11/97

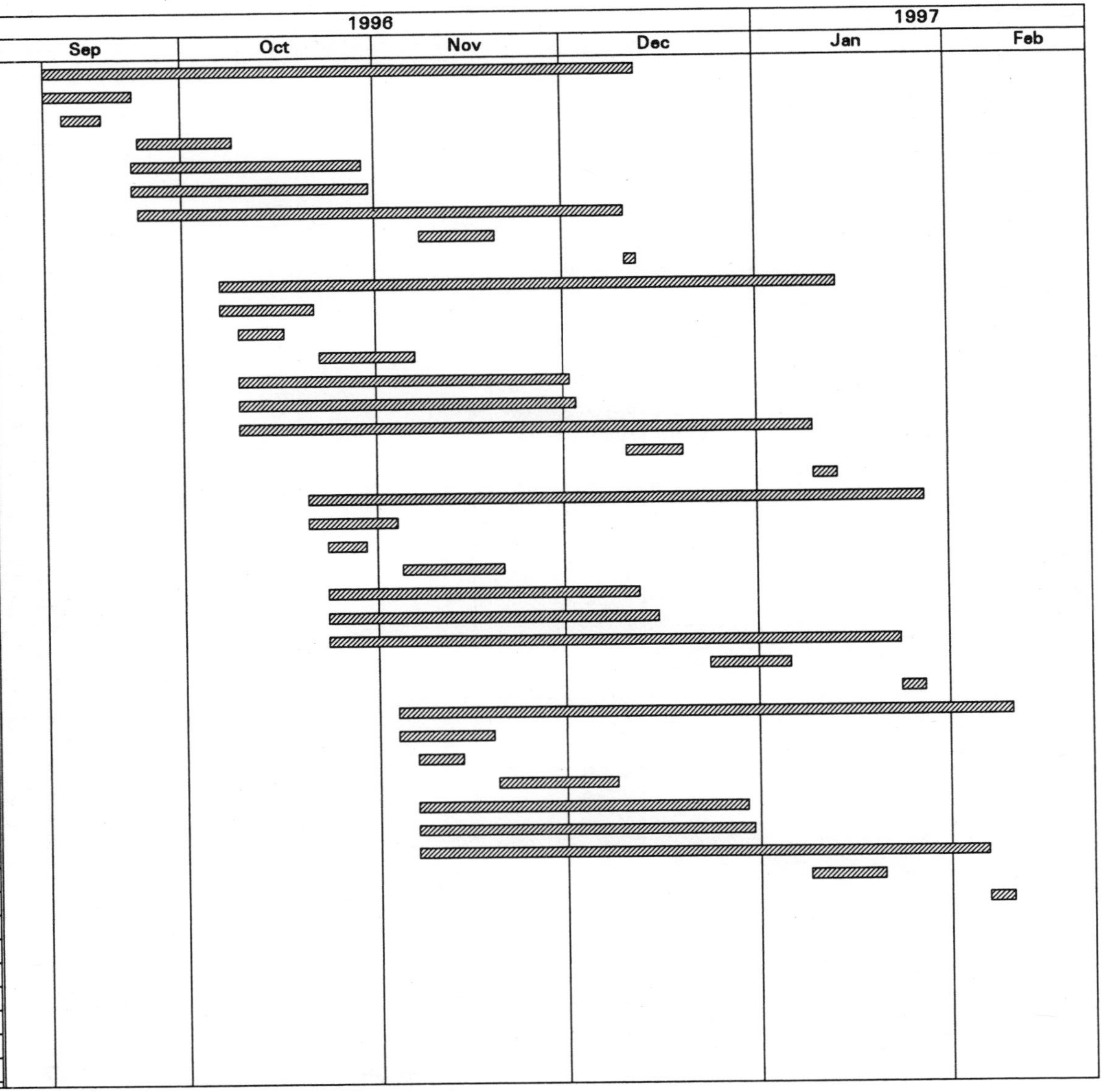

1996: Sep, Oct, Nov, Dec — 1997: Jan, Feb

Figure 4.14: Master Schedule Showing Electrical Sub Activities

Project/Phase/Activity List	Resp	Duration	Start	End
The House 1 Schedule		**65.50 d**	**Sep/09/96**	**Dec/13/96**
Project Startup Phase		11.00 d	Sep/09/96	Sep/23/96
Obtain Temporary Power	Electrician	2.00 d	Sep/12/96	Sep/13/96
Structure Phase		26.50 d	Sep/23/96	Oct/30/96
Electrical Rough-in	Electrician	2.00 d	Oct/23/96	Oct/24/96
Electrical Inspection	Electrician	0.50 d	Oct/25/96	Oct/25/96
Finish Phase		52.50 d	Sep/24/96	Dec/11/96
Electrical Trim-out	Electrician	1.00 d	Dec/06/96	Dec/09/96
Final Electrical Inspection	Electrician	0.50 d	Dec/09/96	Dec/09/96
The House 2 Schedule		**65.50 d**	**Oct/07/96**	**Jan/14/97**
Project Startup Phase		11.00 d	Oct/07/96	Oct/22/96
Obtain Temporary Power	Electrician	2.00 d	Oct/10/96	Oct/11/96
Structure Phase		33.50 d	Oct/10/96	Dec/02/96
Electrical Rough-in	Electrician	2.00 d	Nov/21/96	Nov/22/96
Electrical Inspection	Electrician	0.50 d	Nov/25/96	Nov/25/96
Finish Phase		60.50 d	Oct/10/96	Jan/10/97
Electrical Trim-out	Electrician	1.00 d	Jan/07/97	Jan/08/97
Final Electrical Inspection	Electrician	0.50 d	Jan/08/97	Jan/08/97
The House 3 Schedule		**65.50 d**	**Oct/21/96**	**Jan/28/97**
Project Startup Phase		11.00 d	Oct/21/96	Nov/04/96
Obtain Temporary Power	Electrician	2.00 d	Oct/24/96	Oct/25/96
Structure Phase		33.50 d	Oct/24/96	Dec/13/96
Electrical Rough-in	Electrician	2.00 d	Dec/06/96	Dec/09/96
Electrical Inspection	Electrician	0.50 d	Dec/10/96	Dec/10/96
Finish Phase		60.50 d	Oct/24/96	Jan/24/97
Electrical Trim-out	Electrician	1.00 d	Jan/21/97	Jan/22/97
Final Electrical Inspection	Electrician	0.50 d	Jan/22/97	Jan/22/97
The House 4 Schedule		**65.50 d**	**Nov/04/96**	**Feb/11/97**
Project Startup Phase		11.00 d	Nov/04/96	Nov/19/96
Obtain Temporary Power	Electrician	2.00 d	Nov/07/96	Nov/08/96
Structure Phase		33.50 d	Nov/07/96	Dec/30/96
Electrical Rough-in	Electrician	2.00 d	Dec/20/96	Dec/23/96
Electrical Inspection	Electrician	0.50 d	Dec/24/96	Dec/24/96
Finish Phase		60.50 d	Nov/07/96	Feb/07/97
Electrical Trim-out	Electrician	1.00 d	Feb/04/97	Feb/05/97
Final Electrical Inspection	Electrician	0.50 d	Feb/05/97	Feb/05/97

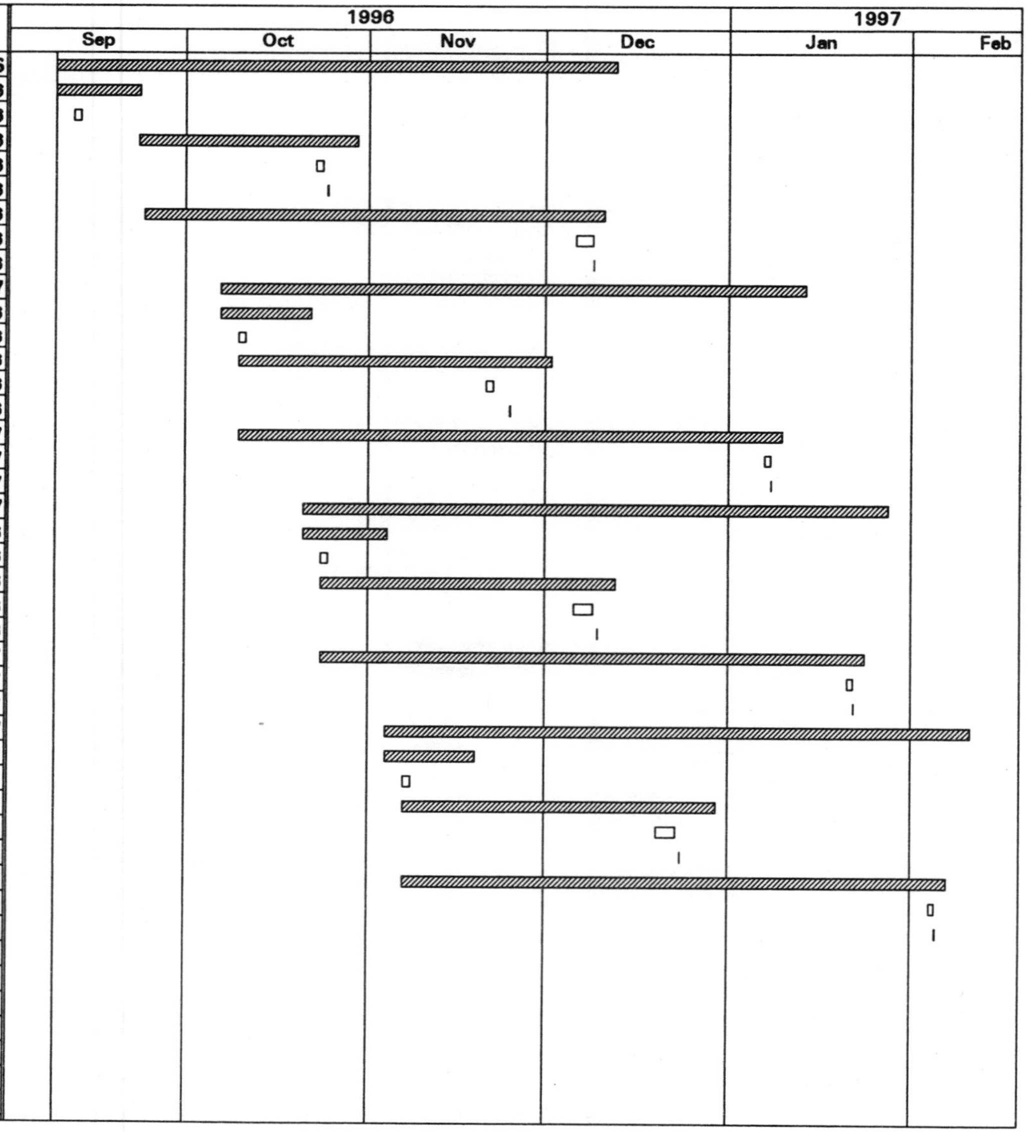

1996 1997

Sep Oct Nov Dec Jan Feb

Figure 4.15: Master Schedule Showing Framing Sub Activities

Project/Phase/Activity List	Resp	Duration	Start	End
The House 1 Schedule		**65.50 d**	**Sep/09/96**	**Dec/13/96**
Structure Phase		26.50 d	Sep/23/96	Oct/30/96
Wall and Sheathing	Framing	3.00 d	Oct/03/96	Oct/07/96
Roof Framing & Decking	Framing	5.00 d	Oct/10/96	Oct/17/96
Install Deadwood and Blk	Framing	0.50 d	Oct/18/96	Oct/18/96
Structure Dry-in Phase		27.50 d	Sep/23/96	Oct/31/96
Install Fascia and Soffits	Framing	4.00 d	Oct/18/96	Oct/24/96
Install Ext Doors & Win	Framing	4.00 d	Oct/18/96	Oct/24/96
Install Overhead Door	Framing	0.50 d	Oct/24/96	Oct/24/96
Install Siding	Framing	0.50 d	Oct/25/96	Oct/25/96
The House 2 Schedule		**65.50 d**	**Oct/07/96**	**Jan/14/97**
Structure Phase		33.50 d	Oct/10/96	Dec/02/96
Wall and Sheathing	Framing	3.00 d	Nov/01/96	Nov/05/96
Roof Framing & Decking	Framing	5.00 d	Nov/08/96	Nov/15/96
Install Deadwood and Blk	Framing	0.50 d	Nov/18/96	Nov/18/96
Structure Dry-in Phase		34.50 d	Oct/10/96	Dec/03/96
Install Fascia and Soffits	Framing	4.00 d	Nov/18/96	Nov/22/96
Install Ext Doors & Win	Framing	4.00 d	Nov/18/96	Nov/22/96
Install Overhead Door	Framing	0.50 d	Nov/22/96	Nov/22/96
Install Siding	Framing	0.50 d	Nov/25/96	Nov/25/96
The House 3 Schedule		**65.50 d**	**Oct/21/96**	**Jan/28/97**
Structure Phase		33.50 d	Oct/24/96	Dec/13/96
Wall and Sheathing	Framing	3.00 d	Nov/15/96	Nov/19/96
Roof Framing & Decking	Framing	5.00 d	Nov/22/96	Dec/02/96
Install Deadwood and Blk	Framing	0.50 d	Dec/03/96	Dec/03/96
Structure Dry-in Phase		34.50 d	Oct/24/96	Dec/16/96
Install Fascia and Soffits	Framing	4.00 d	Dec/03/96	Dec/09/96
Install Ext Doors & Win	Framing	4.00 d	Dec/03/96	Dec/09/96
Install Overhead Door	Framing	0.50 d	Dec/09/96	Dec/09/96
Install Siding	Framing	0.50 d	Dec/10/96	Dec/10/96
The House 4 Schedule		**65.50 d**	**Nov/04/96**	**Feb/11/97**
Structure Phase		33.50 d	Nov/07/96	Dec/30/96
Wall and Sheathing	Framing	3.00 d	Dec/03/96	Dec/05/96
Roof Framing & Decking	Framing	5.00 d	Dec/10/96	Dec/16/96
Install Deadwood and Blk	Framing	0.50 d	Dec/17/96	Dec/17/96
Structure Dry-in Phase		34.50 d	Nov/07/96	Dec/31/96
Install Fascia and Soffits	Framing	4.00 d	Dec/17/96	Dec/23/96
Install Ext Doors & Win	Framing	4.00 d	Dec/17/96	Dec/23/96
Install Overhead Door	Framing	0.50 d	Dec/23/96	Dec/23/96
Install Siding	Framing	0.50 d	Dec/24/96	Dec/24/96

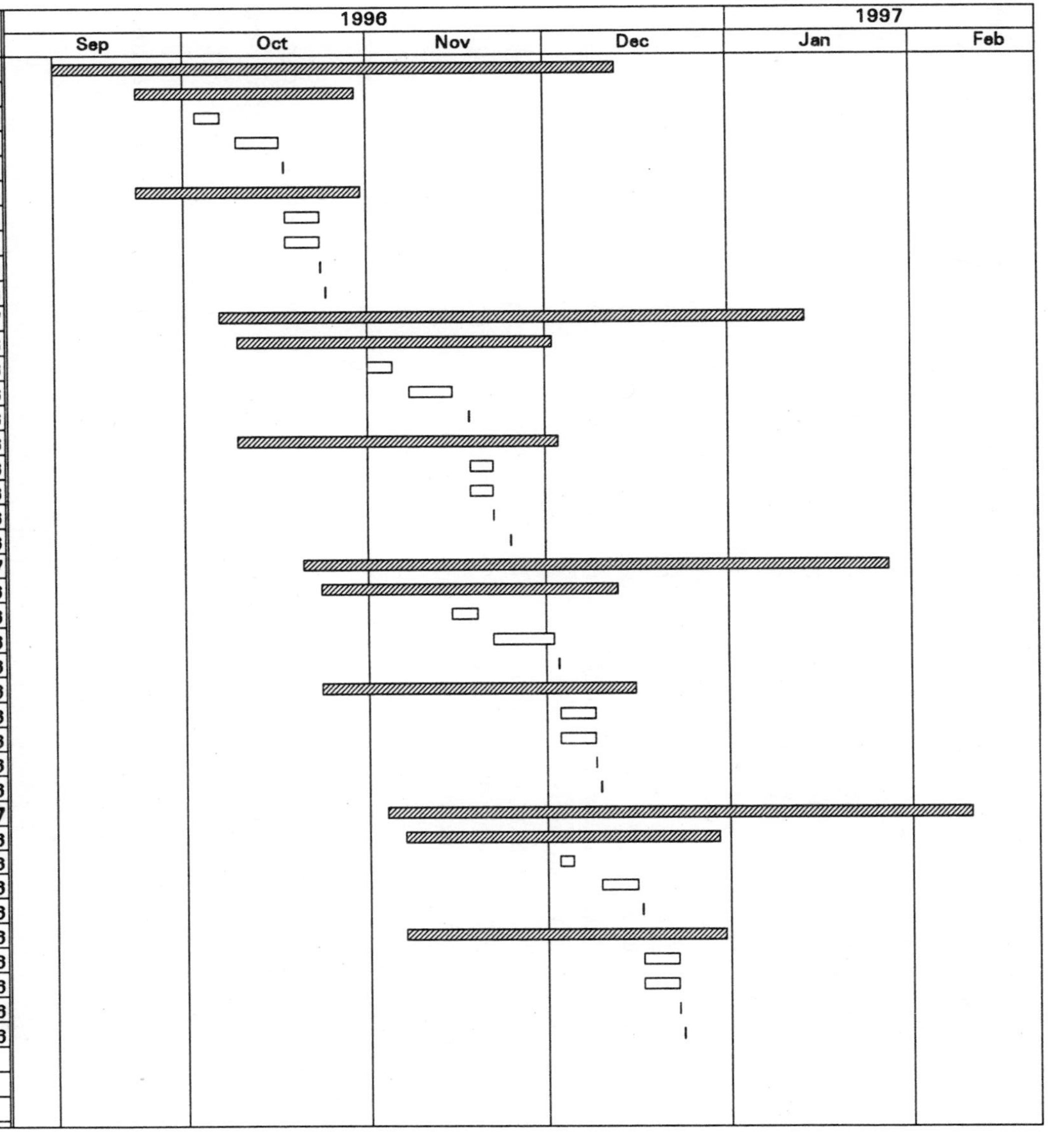

SUMMARY POINTS AND TIPS

- Good communication is essential to completing your projects on time. Builders or remodelers can use the project schedule as a tool to communicate with employees, subcontractors, owners, and lenders.
- As a tool for tracking job progress, the schedule is updated and revised to reflect the current status. Updating the schedule completes the planning cycle by incorporating necessary adjustments into future schedules.
- Builders and remodelers enhance their project control by using the schedule to plan for future events, adjust resource needs, and verify subcontractor resource requirements.
- Cash flow can be monitored in conjunction with the schedules. Computerization helps with producing cash-flow diagrams. Remember that the numbers are rough estimates based on estimated costs and estimated time requirements.
- The schedule can help the builder or remodeler solve resource conflicts that arise in the course of construction.
- When delays occur, the schedule can be crashed by reducing duration times, adding workers, and using other techniques. In considering alternatives to get a schedule back on track, builders or remodelers should start first with those critical activities that are under their direct control and that will make the most difference in recovering the lost time. Good communication and teamwork are the essential components of any successful attempt to adjust the schedule.
- Multiproject scheduling provides an overall picture of company activities. It can provide information that helps the builder or remodeler plan for future work. Company cash flow can be anticipated and managed, and resources allocated efficiently to help the company save money.
- Scheduling also can be used to help builders or remodelers educate subcontractors to better manage their resources.

Chapter 5

Scheduling With Computers

This chapter presents some of the advantages of using computerized scheduling software. In recent years, tremendous improvements have been made in computer hardware and software.

Current computer technology offers builders and remodelers two advantages: it reduces the time required to generate the schedule and the associated reports, graphics, and analysis; and it improves communication by generating custom reports for specific people such as subcontractors, owners, and bankers.

Builders and remodelers underuse computerized scheduling for three reasons: their failure to obtain proper education and training, inadequate computer hardware or software, and their lack of commitment to *use* the scheduling information.

As a business tool, the computer and its software can become as important to you as any construction equipment. It cannot replace your understanding of the construction process or your ability to coordinate subcontractors, but it can greatly enhance how you apply those skills.

Many people think the computer can do it all. In fact, the computer can only do what it is told. You must determine what activities are to be included in the schedule, the logic of how those activities fit together, and the duration to be assigned to each activity. The computer will then perform the network calculations, sort information as requested, and generate graphs and reports quickly and accurately.

HARDWARE

One important question to ask yourself when purchasing computer hardware is: "How much do I want to spend?" Changes in the computer industry have increased both the options and the confusion associated with the purchase of computer systems. Processor types, hard drive sizes, video capabilities, amount of RAM (memory), CD-ROM disk drives, and operating systems are all variables that must be considered.

Hardware that is state-of-the-art today can be completely outdated within two to three years. This is not to say that the equipment will be unusable. It does mean that at some point the latest software updates require a faster, newer system. You will have to judge whether your needs are adequately met by your current system or whether you should upgrade your hardware to take advantage of the latest software developments.

Some computer sales associates advise builders to select their software first, then buy hardware that can handle the desired software. If you do this, you should consider *all* the software that you might need, such as accounting, spreadsheet, estimating, and computer-aided design software. Builders or remodelers who want to add scheduling programs to their existing systems need to verify that the software will perform adequately on their current system. The amount of RAM, type of printer, available hard disk space, and operating system will be some of the factors to check. If possible, get a sample program or a trial copy of the software to check its performance on your system. This is the only true test.

So what type of computer hardware do you purchase? This question can only be answered at the time the purchase is made. What is required today may not be adequate for tomorrow. Questions such as, "Do I need a color monitor?" or "Do I need a laser printer or will a bubble jet be able to handle the kinds of graphics I want to generate?" cannot be answered today. Ask fellow builders or remodelers who are using scheduling software for their advice, or contact the NAHB Software Review Program for current information about software being used in the industry. Many commercial publications report on software and hardware developments, including *Fortune*, *PC Computing*, and *Consumer Reports*.

Remember that whatever you purchase likely will be less than state-of-the-art in a fairly short time. Budget for a system or upgrade that will meet your anticipated needs for two to three years. At the end of this period, re-evaluate the system. At its most basic level, your decision needs to be based on the question: "Will this system do what I need it to do for the money I have to spend?"

SOFTWARE

Scheduling software is available at a wide range of prices. Basic software can cost as little as $100, while highly advanced packages can cost more than $4,000. Many of the most advanced packages have features that the small-to-medium volume builder can live without; however, the least expensive software may lack important features or support services. The goal is to find software in your price range that has features to fit your needs.

Look for a program that:

- Works well on the hardware you have or plan to acquire
- Provides easy-to-understand methods for entering activity information and for creating and displaying reports and graphs
- Can include activity information such as responsibility, cost, or other special information
- Prints time-scaled bar charts
- Highlights or otherwise identifies resource conflicts or overloads when multiple projects are analyzed together
- (For remodelers) allows durations to be tracked in terms of hours
- Allows the builder or remodeler to track separate workday calendars for each project
- Provides sorting features for selecting specific activities for reports and graphics, for example, according to subcontractor
- Offers a summary feature to compress the activities, so that you can easily change the level of detail you are monitoring
- Is easy to use and easy to understand

Specific features will vary, and you should evaluate several different programs before making a purchase. In recent years, software programs using Windows or other graphic interfaces have made advances in ease of use. They provide many short-cut buttons, look good, and work smoothly, giving the user greater control and providing all information on screen. However, Windows ideally requires hardware with minimum memory capacity of four megabytes (eight megabytes is already recommended and a sixteen-megabyte minimum is coming in the near future) and a faster processor. On older machines, the DOS-based programs may still perform more efficiently. Apple Macintosh computers also can carry scheduling software and provide excellent performance.

Purchasing software with easy-to-understand features will help minimize your training time and keep your learning curve as short as possible. However, you should always plan to commit to some formal training, especially if you have no experience in scheduling.

When evaluating training opportunities, check to see if the course covers the specific information you will need. Scheduling programs can perform many functions, but you will derive the most benefit from courses that use construction examples. Colleges and universities are a good place to start looking for training. Independent consultants also provide training; however, you should verify their claims—ideally by talking with builders or remodelers who have been through their training—before you attend.

Software vendors who provide training may or may not understand construction. A vendor can do a great job of demonstrating a program, but do not let a single demonstration convince you to purchase. If possible, ask for some hands-on time with the software and try to use it to solve a scheduling problem you have experienced. Watching a vendor or trainer perform an operation and performing it yourself are different things. It's better not to purchase something than to purchase something that you will never use. NAHB's Software Review Program and the Home Builder Bookstore can provide information to help you negotiate the process of buying and setting up a computerized scheduling system. For more information, see the list at the back of the book.

Using the Software

Begin by scheduling only one project on the computer system. This first schedule should be prepared from the activities selected from your master list. Be sure to involve subcontractors and employees in the scheduling process, and review the schedule on a regular basis. When you have successfully completed a project using the scheduling software, you may then add additional projects with confidence.

Update this first, or pilot, schedule every week to every two weeks. Updating more frequently tends to create unnecessary work, but a longer time frame may not provide enough warning if problems start to occur. Of course, if the pilot project encounters scheduling problems, more frequent updating may be necessary to help determine the cause of the problem.

The purpose of computerized scheduling is to save time and increase your options—not to create a lot of additional work. A computerized schedule system can:

- Forecast completion dates
- Record and diagram an initial schedule based on preset constraints
- Recalculate schedule dates based on adjustments made in response to contingencies
- Highlight areas in the schedule where changes can be made to offset delays
- Adjust the diagrams to reflect new logic as changes are made
- Record data and generate narrative reports such as activity descriptions, sequence numbers, and special codes, duration estimates, early start and early finish dates (by calendar or by workdays), late start and late finish dates (by calendar or by workdays), actual start and actual finish dates, float/critical path listing, percentage of completion, cost estimates based on projected or cost per unit, comparison of actual dates to the schedule dates, duration, and so forth
- Produce time-scaled bar charts
- Track the original schedule (as a baseline) against the updated schedule for comparisons

- List resource conflicts and suggest possible solutions

A manual scheduling system can produce all the information listed above. However, it would take a tremendous amount of time and energy that could be used to perform more important tasks. Good scheduling requires time. Computer scheduling allows you to get maximum benefit from the time invested.

All of the charts in Chapter 4 were easily created using computer software. Figure 5.1 presents a typical computerized schedule report that might be created for our sample house.

The report contains a lot of data. It would take anyone several hours to unravel the information to perform a complete analysis. Reports of this type can be valuable once a builder has a complete understanding of scheduling methods. However, the typical home builder can gain the same information from a time-scaled bar chart. A graph can replace a lot of written data and can be understood by more people. The computer can produce both types of reports quickly and easily.

One piece of valuable information in this report is the Float column. This is the amount of slippage an activity has available. If an activity has zero float, then the activity is automatically critical. Each critical activity has to be completed by its scheduled time or the delay will affect the completion of the project.

By using the sort function, we can use the computer to generate another report that shows only those activities with zero float (see Figure 5.2).

Is the number of critical activities important? In the case of our sample schedule, most of the project will have to be done on schedule. Any slippage will cause delays in the project. Schedules that contain a large number of critical activities have to be closely monitored to catch emerging problems.

A builder or remodeler can use the information in reports like Figure 5.2 to help control the progress of their projects. Because subcontractors and employees need to focus on when to complete their work rather than on how much float time is in the schedule, reports like Figure 5.2 should not be distributed to subcontractors but should be kept in-house.

INTEGRATED COMPUTER SYSTEMS

As desktop computers have become more powerful, software programmers have been able to create programs with the ability to transfer information from one application or program to another. This feature—sometimes called *integration*—is important for the builder or remodeler who is trying to automate accounting, estimating, and scheduling.

Some software companies sell integrated packages that come already set up for the programs to send information to each other. If you will use an assortment of programs or if you are adding scheduling to existing programs that can be integrated, ask what pieces connect together and how and who sells and supports compatible programs. Talk to people who use integrated programs and packages to get a realistic idea of how well such a system might work for you.

The main idea behind integrated computer software is that you can take an estimate (which has identified the materials required, their cost, and the labor needed to install the materials) from an estimating program and link this information with the activities being tracked in the scheduling program. The estimating information also can be transferred, without retyping, into the accounting system, where purchase orders and subcontracts can be created from the information. Again without retyping, a job costing program can then

Figure 5.1: Typical Schedule Report

Project/Phase List	Resp	Duration	Early Start	Early Finish	Late Start	Late Finish	Float	Cost
The House 1 Schedule		65.50 d	Sep/09/96	Dec/13/96	Sep/09/96	Dec/13/96	0.00 d	$129,217.00
Project Startup Phase		11.00 d	Sep/09/96	Sep/23/96	Sep/09/96	Sep/24/96	1.00 d	$825.00
Sitework Phase		5.00 d	Sep/12/96	Sep/18/96	Sep/12/96	Sep/18/96	0.00 d	$2,700.00
Slab-on-Grade Phase		12.00 d	Sep/24/96	Oct/09/96	Sep/24/96	Oct/09/96	0.00 d	$14,086.00
Structure Phase		26.50 d	Sep/23/96	Oct/30/96	Sep/30/96	Oct/30/96	0.00 d	$30,660.00
Order Framing Material	Builder	2.00 d	Sep/27/96	Sep/30/96	Oct/02/96	Oct/03/96	3.00 d	$0.00
Deliver Framing Material	Builder	1.00 d	Oct/01/96	Oct/01/96	Oct/04/96	Oct/04/96	3.00 d	$7,111.00
Wall and Sheathing	Framing	3.00 d	Oct/03/96	Oct/07/96	Oct/07/96	Oct/09/96	2.00 d	$2,549.00
Roof Framing & Decking	Framing	5.00 d	Oct/10/96	Oct/17/96	Oct/10/96	Oct/17/96	0.00 d	$4,197.00
Select Whirlpool Tub	Owner	1.00 d	Sep/23/96	Sep/23/96	Sep/30/96	Sep/30/96	5.00 d	$0.00
Order & Deliver Tubs	Builder	4.00 d	Sep/24/96	Sep/27/96	Oct/01/96	Oct/04/96	5.00 d	$2,500.00
Plumbing Top-out	Plumber	4.00 d	Oct/18/96	Oct/23/96	Oct/21/96	Oct/24/96	1.00 d	$2,013.00
Plumbing Test & Insp	Plumber	0.50 d	Oct/24/96	Oct/24/96	Oct/25/96	Oct/25/96	1.00 d	$0.00
HVAC Rough-in	AC Sub	3.00 d	Oct/18/96	Oct/22/96	Oct/18/96	Oct/23/96	0.50 d	$3,347.00
Radon System Top-out	Radon Sub	1.00 d	Oct/25/96	Oct/25/96	Oct/25/96	Oct/28/96	0.50 d	$500.00
Install Fireplace	Fire Sub	1.00 d	Oct/08/96	Oct/08/96	Oct/17/96	Oct/17/96	6.00 d	$3,076.00
Electrical Rough-in	Electrician	2.00 d	Oct/23/96	Oct/24/96	Oct/23/96	Oct/25/96	0.50 d	$4,181.00
Electrical Inspection	Electrician	0.50 d	Oct/25/96	Oct/25/96	Oct/28/96	Oct/28/96	1.00 d	$0.00
Install Deadwood and Blk	Framing	0.50 d	Oct/18/96	Oct/18/96	Oct/28/96	Oct/28/96	6.00 d	$150.00
Check Plumb & Square	Builder	0.50 d	Oct/18/96	Oct/18/96	Oct/18/96	Oct/18/96	0.00 d	$150.00
Framing Inspection	Builder	0.50 d	Oct/18/96	Oct/18/96	Oct/28/96	Oct/28/96	6.00 d	$0.00
Clean up All Material	Builder	0.50 d	Oct/28/96	Oct/28/96	Oct/28/96	Oct/28/96	0.50 d	$50.00
City Inspection	Inspector	0.50 d	Oct/28/96	Oct/28/96	Oct/29/96	Oct/29/96	0.50 d	$0.00
Wall Insulation	Insulator	1.00 d	Oct/29/96	Oct/30/96	Oct/29/96	Oct/30/96	0.00 d	$836.00
Structure Dry-in Phase		27.50 d	Sep/23/96	Oct/31/96	Oct/08/96	Dec/13/96	28.00 d	$21,145.00
Finish Phase		52.50 d	Sep/24/96	Dec/11/96	Oct/04/96	Dec/13/96	2.00 d	$48,472.00
Finish Sitework Phase		7.50 d	Nov/08/96	Nov/20/96	Dec/03/96	Dec/12/96	13.50 d	$8,600.00
Project Close-out Phase		2.00 d	Dec/11/96	Dec/13/96	Dec/11/96	Dec/13/96	0.00 d	$500.00

Note: We have expanded only the Structure Phase to show individual activities.

Shaded activities are on the critical path.

Figure 5.2: Critical Activities Schedule Report

Project/Phase/Activity List	Resp.	Duration	Early Start	Early Finish	Late Start	Late Finish	Float	Cost
The House 1 Schedule		65.50 d	Sep/09/96	Dec/13/96	Sep/09/96	Dec/13/96	0.00	$126,788.00
Project Startup Phase		11.00 d	Sep/09/96	Sep/23/96	Sep/09/96	Sep/24/96	1.00	$825.00
Obtain Permit	Builder	3.00 d	Sep/09/96	Sep/11/96	Sep/09/96	Sep/11/96	0.00	$500.00
Surv Locate Bldg Corners	Surveyor	3.00 d	Sep/19/96	Sep/23/96	Sep/19/96	Sep/23/96	0.00	$250.00
Sitework Phase		5.00 d	Sep/12/96	Sep/18/96	Sep/12/96	Sep/18/96	0.00	$2,700.00
Site Grading	Site Sub	5.00 d	Sep/12/96	Sep/18/96	Sep/12/96	Sep/18/96	0.00	$2,700.00
Slab-on-Grade Phase		12.00 d	Sep/24/96	Oct/09/96	Sep/24/96	Oct/09/96	0.00	$14,086.00
Lay Out & Set Batterboard	Builder	1.00 d	Sep/24/96	Sep/24/96	Sep/24/96	Sep/24/96	0.00	$350.00
Spread Gravel	Builder	0.50 d	Sep/30/96	Sep/30/96	Sep/30/96	Sep/30/96	0.00	$552.00
Plumbing Rough-in	Plumber	3.00 d	Sep/25/96	Sep/27/96	Sep/25/96	Sep/27/96	0.00	$2,175.00
Termite Treatment	Termite	0.50 d	Sep/30/96	Sep/30/96	Sep/30/96	Sep/30/96	0.00	$1,000.00
Set Grade Stakes	Builder	0.50 d	Oct/01/96	Oct/01/96	Oct/01/96	Oct/01/96	0.00	$50.00
Place Poly and WWM	Builder	0.50 d	Oct/01/96	Oct/01/96	Oct/01/96	Oct/01/96	0.00	$1,056.00
Inspect Slab	Inspector	0.50 d	Oct/01/96	Oct/01/96	Oct/01/96	Oct/01/96	0.00	$0.00
Place & Finish Concrete	Builder	1.00 d	Oct/02/96	Oct/02/96	Oct/02/96	Oct/02/96	0.00	$3,952.00
Place Garage/Storage Slab	Builder	3.00 d	Oct/03/96	Oct/07/96	Oct/03/96	Oct/07/96	0.00	$1,934.00
Place Concrete Patio	Builder	2.00 d	Oct/08/96	Oct/09/96	Oct/08/96	Oct/09/96	0.00	$925.00
Structure Phase		26.50 d	Sep/23/96	Oct/30/96	Sep/30/96	Oct/30/96	0.00	$30,660.00
Roof Framing & Decking	Framing	5.00 d	Oct/10/96	Oct/17/96	Oct/10/96	Oct/17/96	0.00	$4,197.00
Check Plumb & Square	Builder	0.50 d	Oct/18/96	Oct/18/96	Oct/18/96	Oct/18/96	0.00	$150.00
Wall Insulation	Insulator	1.00 d	Oct/29/96	Oct/30/96	Oct/29/96	Oct/30/96	0.00	$836.00
Structure Dry-in Phase		27.50 d	Sep/23/96	Oct/31/96	Oct/08/96	Dec/13/96	28.00	$21,145.00
Install Fascia and Soffits	Framing	4.00 d	Oct/18/96	Oct/24/96	Oct/18/96	Oct/24/96	0.00	$1,871.00
Install Shingles	Roofer	3.00 d	Oct/24/96	Oct/29/96	Oct/24/96	Oct/29/96	0.00	$3,741.00
Finish Phase		52.50 d	Sep/24/96	Dec/11/96	Oct/04/96	Dec/13/96	2.00	$48,472.00
Hang Sheetrock	Drywall Sub	2.00 d	Oct/30/96	Nov/01/96	Oct/30/96	Nov/01/96	0.00	$500.00
Tape & Finish Sheetrock	Drywall Sub	4.00 d	Nov/01/96	Nov/07/96	Nov/01/96	Nov/07/96	0.00	$1,444.00
Clean up Sheetrock Waste	Drywall Sub	1.00 d	Nov/07/96	Nov/08/96	Nov/07/96	Nov/08/96	0.00	$200.00
Install Interior Doors	Trim Sub	1.00 d	Nov/08/96	Nov/12/96	Nov/08/96	Nov/12/96	0.00	$667.00
Install Interior Trim	Trim Sub	2.00 d	Nov/12/96	Nov/14/96	Nov/12/96	Nov/14/96	0.00	$848.00
Paint Interior	Painter	4.00 d	Nov/14/96	Nov/20/96	Nov/14/96	Nov/20/96	0.00	$1,970.00

Figure 5.2: Critical Activities Schedule Report, Continued

Project/Phase/Activity List	Resp.	Duration	Early Start	Early Finish	Late Start	Late Finish	Float	Cost
Install Cabinets	Cab Sub	3.00 d	Nov/26/96	Dec/03/96	Nov/26/96	Dec/03/96	0.00	$1,580.00
Deliver & Set Appliances	Builder	1.00 d	Dec/10/96	Dec/11/96	Dec/10/96	Dec/11/96	0.00	$3,735.00
Hang Wallpaper	Painter	4.00 d	Nov/20/96	Nov/26/96	Nov/20/96	Nov/26/96	0.00	$1,140.00
Install Clay Tile	Floor Sub	5.00 d	Dec/03/96	Dec/10/96	Dec/03/96	Dec/10/96	0.00	$4,448.00
Install Ceramic Tile	Floor Sub	2.00 d	Dec/03/96	Dec/05/96	Dec/03/96	Dec/05/96	0.00	$1,135.00
Plumbing Trim-out	Plumber	1.00 d	Dec/05/96	Dec/06/96	Dec/05/96	Dec/06/96	0.00	$2,314.00
Install Marble	Floor Sub	2.00 d	Dec/09/96	Dec/11/96	Dec/09/96	Dec/11/96	0.00	$200.00
Install Carpet	Floor Sub	2.00 d	Dec/09/96	Dec/11/96	Dec/09/96	Dec/11/96	0.00	$2,001.00
Electrical Trim-out	Electrician	1.00 d	Dec/06/96	Dec/09/96	Dec/06/96	Dec/09/96	0.00	$4,137.00
Project Close-out Phase		2.00 d	Dec/11/96	Dec/13/96	Dec/11/96	Dec/13/96	0.00	$300.00
Paint Touch-up	Painter	0.50 d	Dec/11/96	Dec/11/96	Dec/11/96	Dec/11/96	0.00	$200.00
Clean House	Builder	0.50 d	Dec/12/96	Dec/12/96	Dec/12/96	Dec/12/96	0.00	$50.00
Complete Punch List	Builder	0.50 d	Dec/12/96	Dec/12/96	Dec/12/96	Dec/12/96	0.00	$50.00
Final Inspection	Builder	0.50 d	Dec/13/96	Dec/13/96	Dec/13/96	Dec/13/96	0.00	$0.00

be used to track the invoices to determine budget variances.

Many builders already use integrated estimating and accounting programs. Scheduling integration is now available and is starting to be used. The integration will increase the use of cash-flow analysis and resource tracking procedures in schedules.

Manual methods of sorting and collecting cost information in relation to the schedule can be a burden. Using integrated estimating, scheduling, and accounting programs, the builder or remodeler can easily add cost loading and resource tracking to their schedules without expending a lot of extra time.

However, the use of integrated software is not always an easy task. The builder, remodeler, or staff person using the system must have sufficient training to operate multiple programs. For an inexperienced computer user, such juggling can seem like a nightmare. Integration also amplifies the "Garbage-in, Garbage-out" theory. If the basic information entered into the system is faulty, then any of the reports generated using that information will be flawed. Also, when using integrated software, the importance of keeping information current escalates. If information in one area is not current, this can affect other areas. For example, can payments for invoices be generated from the accounting program if deliveries have not been marked complete in the schedule?

Integration can be a time-saving tool when applied correctly. It takes computer understanding, program understanding, and confidence in one's own information. Start slow and build a system on sound information. Add advanced features as your understanding progresses and the problems will be minimal.

SUMMARY POINTS AND TIPS

- Hardware and software improvements have made computer scheduling affordable and easier to operate.
- The initial purchase of the hardware and software is a small cost when compared to the training and operating commitment that is required to make use of the system.
- Using computers can reduce the time spent drawing schedules and calculating completion times of projects, but it will not replace the builder's understanding and knowledge. The builder's or remodeler's plan for completing the work will still be the most important part of the system.
- Before purchasing software, ask for a hands-on test—ideally on your computer—and make sure the program addresses the things *you* need done. Unnecessary "bells and whistles" will not compensate for a basic mismatch between the program and your needs.
- Consider purchasing an integrated computer system. Such systems can reduce the time spent rekeying information by automatically linking processes such as estimating, purchasing, and scheduling. Integrated systems require more computer understanding, however, and a poorly designed system can cause performance problems.

Chapter 6

Scheduling a Remodeling Project

Remodelers have special considerations when applying scheduling techniques to their projects, including:

- Working part days
- Working in areas with finished work
- Working with occupants living in areas to be remodeled
- Working with tight time frames to remove and improve areas in use

Successful remodelers handle these and other problems with a minimum of stress. This chapter will present ways to address these issues and create schedules to help reduce problems.

A remodeler can apply all of the scheduling techniques discussed in the previous chapters. The nature of a remodeling project can give the schedule more immediacy, however. Remodeling work typically follows three stages: demolition, reconstruction, and new construction.

The *demolition* stage consists of the removal of any existing construction. This stage can be as simple as removing a section of flooring or as complex as removing an entire roof system. Demolition usually precedes the other stages of a remodeling job. However, for larger jobs, demolition may itself be scheduled in phases.

The *reconstruction* stage encompasses the installation of the new materials. The remodeler who has prepared a complete plan and a workable schedule for this stage will avoid many customer relations problems. This stage follows demolition as soon as possible.

The *new construction* stage occurs when the remodeling project consists of a freestanding addition. These projects can be scheduled following a process similar to that used for the garage in Chapter 2.

Remodelers must closely coordinate communication among the owner, subcontractors, and the remodeler's own crew. Potential liability for damage to existing work is always present, and time delays are directly felt by the owner.

Depending on the project and project conditions, the remodeler also must manage:

- Working part days. The activities of the homeowner may be such that a limited workday may be required. For example, access to the home may be restricted to the hours from 10:00 a.m. to 2:00 p.m. In the schedule, durations for each activity may need to be adjusted to accommodate these limitations.
- Limited physical access to the property. Use of heavy equipment or vehicles can be limited, and the site may require complete cleanup at the end of each day.

- Demolition "surprises." The remodeler bears the risk of having to alter the schedule based on what may be discovered inside a wall. Also, without careful planning, damage to other parts of the home can occur. These risks make it imperative to allow adequate time during the demolition stage to cover contingencies.
- The presence of the homeowner, delaying work at various stages. Worker performance can be slowed by the presence of other people or by constant owner supervision. Also, the safety of the workers, owners, and others is a consideration.
- Extremely tight time frames. Remodelers must complete their work with as little impact on the homeowner as possible. Despite the constraints and complications of remodeling work, speed is important to both the remodeler and the homeowner.
- Unique scheduling problems. It can be extremely difficult to use manual techniques to solve scheduling problems for a remodeling project. This is because the remodeler has two unique needs: First, the schedule generally is created, calculated, and manipulated in terms of hours instead of days or half days. Second, the calendar for working and non-working days is likely to vary with each project. The number of work-hours per day may also vary. This makes the manual calculation of work-hours and the conversion to calendar days difficult and time-consuming. Remodelers must keep these considerations in mind when shopping for scheduling software.

SAMPLE KITCHEN PROJECT

Our sample remodeling project has several parts. The owner wishes to do some remodeling of existing space, extend a living space on the end of the house, add garage doors, and add a privacy fence. The plans in Appendix C identify the areas to be addressed and Appendix D contains the estimate for this work.

The owner has determined that the first part of this project is to remove the kitchen cabinets and replace them. The new cabinets will cover the floor area in a manner similar to the existing cabinets, except for a new bar counter extending to the end of the island. Some floor and wall rework will need to be done. Existing wallpaper will have to be reworked since the homeowner wants it to remain.

The second task is to add a room by removing the brick from the end wall of the wardrobe and extending the roof for approximately fifteen feet to create a workout room. The roof will have to be reworked and the gable window will move to the end of the house. Existing roofing and rafters will have to be removed and the exterior masonry will be removed and reconnected. Two skylights will replace the gable window in the master bath.

The owner also wishes to have an enclosed private courtyard with a fountain. The opening between the wings of the house will be closed with a six-foot-high wood fence. (The extension for the workout room will make the new fence run perpendicular to each wing.) Extra landscaping will be added to the courtyard area as directed by the owner.

The final adjustment is to close in the garage. The opening in the carport will be closed by adding two ten-foot-wide overhead garage doors. This will seal off the courtyard area and increase the privacy.

THE PLAN

Because this project deals with three areas—kitchen, addition, and exterior—the planning breakdown can vary. We will create a solution for each individual project. Using a computer program, we can create a single plan using the outline feature with headings by area and then by operation.

Another approach requires the use of three separate schedules. This is similar to handling multiple projects for new construction. A specific schedule dealing with each of the three areas can be developed and then combined to see the overall project sequencing. If you prefer this planning approach, be sure your software can easily perform this function.

Because remodeling projects vary so greatly in detail, it is difficult to develop a standard listing. The scope and type of work, existing structure, and relationship with the homeowner present different challenges on each project.

An essential step in developing a remodeling project plan is to discover any special needs or demands of the homeowner. In our remodeling example, this homeowner has agreed to allow the kitchen area to be out of operation for five workdays. The master bath and wardrobe areas can be out of operation for another one-week (five workday) period. These areas may be out of operation at different times. The exterior work can be done at any point during the project. The owner does not have a specific time for completion, except for wanting the job finished as soon as possible.

As in our house example, the first task is to gather necessary information from subcontractors and material suppliers regarding the planned activities. Many of the activities for a project of this type may have short durations, often less than half a day.

Short-duration activities include tasks such as "Turn-off water," "Bring tarp to cover roof," or "Remove flooring for new cabinets." These activities can take only minutes to complete; however, listing them in the plan helps ensure that they are not forgotten. The inconvenience of including these small tasks is more than offset by avoiding the damage or delays that occur if such tasks are overlooked.

We will develop an outline for this job in three separate phases. The final schedule will then be coordinated to meet the owner's time requirements and determine the project completion date.

The Kitchen Schedule

The plan for the kitchen work consists of the activities shown in Figure 6.1. The duration and responsibility are identified after each activity. Some activities are scheduled for less than a half day. Working out the durations in hours provides better time estimates for calculating the completion date. Also, note that the remodeler will perform most of these activities, which means the remodeler has a fair amount of control over this work.

We will schedule each section independently. A diagram of this plan for the kitchen appears in Figure 6.2. This diagram gives the remodeler an initial look at this section of the project. We will assume that the planned start date for this area and project is January 25, 1996. This may change for this area when the schedules are combined later to help organize the completed overall project schedule.

Figure 6.3 provides a more detailed look at the week of March 3, when the remodeler has planned to close the kitchen and complete the work on this phase of the project. The linchpin of this portion of the schedule is the delivery date for the cabinets. Because this is an hour-based schedule, the start and finish days could be removed from the diagram. Many activities start and finish on the same day, and this is reflected on the graph.

The owner selection dates and the order times also control when the kitchen will be renovated. The remodeler has talked to the cabinet subcontractor and determined that a three- to four-week time frame is needed for building and delivery of the cabinets. The cabinet

Figure 6.1: Plan for Kitchen Remodel

Task	Duration, Responsible
The Kitchen	
Demolition and Removal	
Remove Existing Cabinets	6 hrs., Remodeler
Disconnect and Remove Appliances	4 hrs., Remodeler
Disconnect and Remove Sink	1 hr., Remodeler
Remove Floor at New Cabinet	2 hrs., Remodeler
Dispose of Cabinets	2 hrs., Remodeler
Patching and Repair	
Repair Damage to Walls	2 hrs., Remodeler
Repair Damage to Wallpaper	3 hrs., Remodeler
Replace Flooring as Needed	1 day, Remodeler
New Work	
Cabinets	
Owner Select Cabinets	1 day, Owner
Order New Cabinets	1 day, Remodeler
Build and Deliver	24 days, Cabinet Sub
Install New Cabinets	12 hrs., Cabinet Sub
Install New Countertops	4 hrs., Cabinet Sub
Replace Appliances	4 hrs., Remodeler
Install Sink	2 hrs., Plumber

subcontractor agrees that the five-week (twenty-five workday) timeframe we have placed in the schedule will ensure the work can be completed on March 8.

Once the cabinets are delivered, we can begin the interior demolition and removal work. After the existing cabinets and floor are removed, the new cabinets can be installed. Disposal of the existing cabinets can be taken care of during this time as well. The countertops and the repair work to the walls, flooring, and wallpaper can begin after the cabinets. Note that the repair work and the removal work are scheduled one activity at a time. This work could happen in a different sequence and some activities could overlap. The remodeler has to determine the logic that is best for this situation. After the countertops are set, the sink can be replaced. The last activity in this area is the placing of the appliances.

The Addition Schedule

Scheduling work for the addition is a little more complex. The existing structure has to be modified and a new slab, walls, and roof added. We must take care to minimize damage to the existing yard and landscaping when moving material to the new location. We also must determine a way to place the twenty cubic yards of concrete because the truck cannot reach the area for placement. We can use a concrete pump; use wheelbarrows to move the concrete, then mix and pour sacks of concrete onsite; or use conveyor belts provided by the concrete company.

Each of these concrete placement options has advantages and disadvantages. Another solution may present itself that is more cost-effective and that produces the desired result. The remodeler should select the alternative that best minimizes cost and maintains quality.

Figure 6.2: Kitchen Remodel Diagram

Project/Phase/Activity List	Duration	Start	End
The Kitchen	31.00 d	Jan/25/96	Mar/08/96
Demo & Removal	13.00 h	Mar/04/96	Mar/05/96
Dispose of Cabinets	2.00 h	Mar/05/96	Mar/05/96
Remove Existing Cabinets	6.00 h	Mar/04/96	Mar/05/96
Disc and Remove Applianc	4.00 h	Mar/04/96	Mar/04/96
Disc and Remove Sink	1.00 h	Mar/04/96	Mar/04/96
Remove Floor as required	2.00 h	Mar/05/96	Mar/05/96
Patching and Repair	1.63 d	Mar/07/96	Mar/08/96
Replace Flooring	1.00 d	Mar/07/96	Mar/08/96
Repair Wallpaper	3.00 h	Mar/07/96	Mar/07/96
Repair Walls	2.00 h	Mar/07/96	Mar/07/96
New Work	31.00 d	Jan/25/96	Mar/08/96
Owner Select Cabinets	1.00 d	Jan/25/96	Jan/25/96
Order New Cabinets	1.00 d	Jan/26/96	Jan/26/96
Build and Deliver Cabinets	24.00 d	Jan/29/96	Mar/01/96
Install Cabinets	1.50 d	Mar/05/96	Mar/07/96
Install Countertops	4.00 h	Mar/07/96	Mar/07/96
Replace Appliances	2.00 h	Mar/08/96	Mar/08/96
Install Sink	2.00 h	Mar/07/96	Mar/07/96

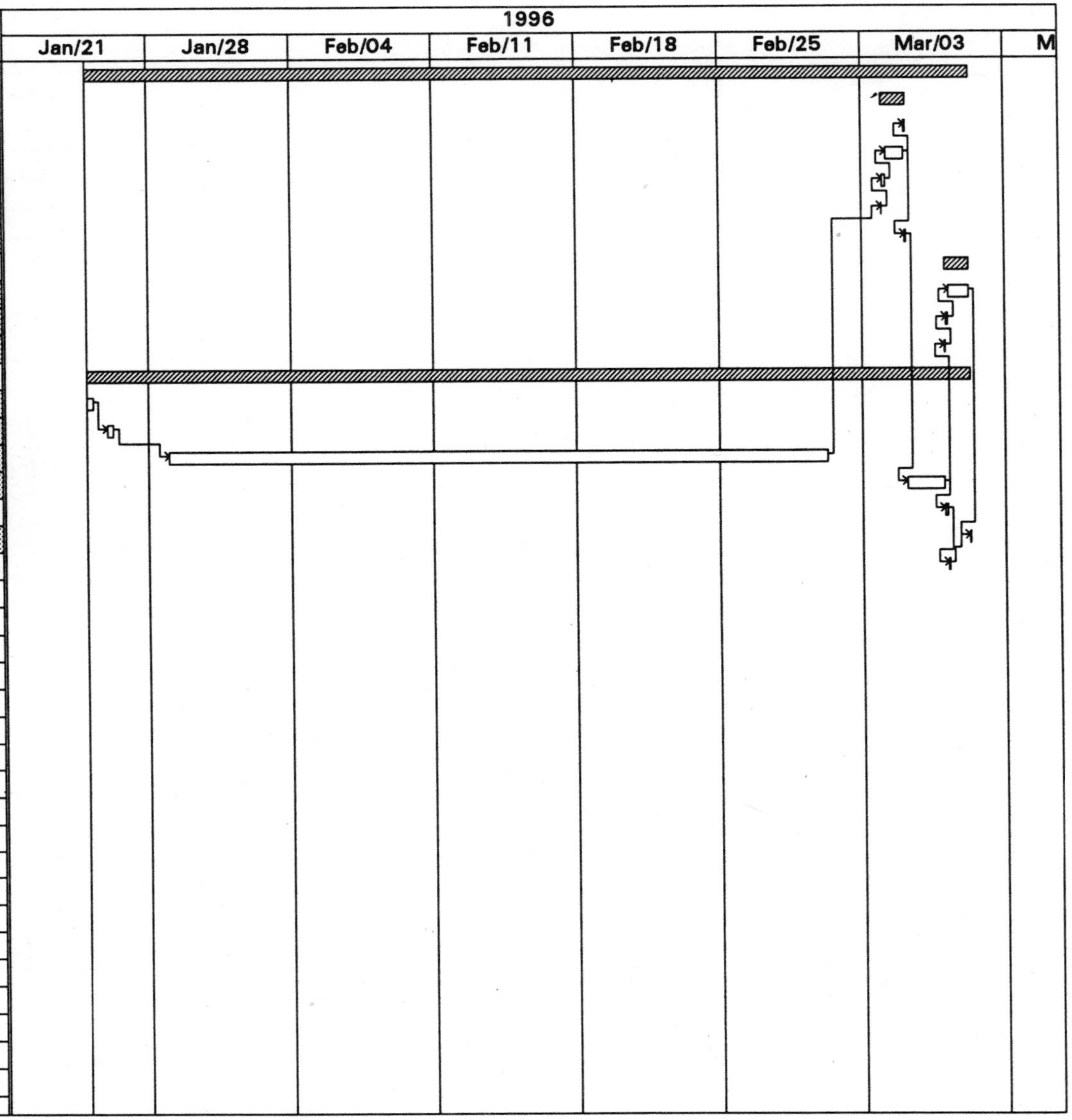

Shaded activities are on the critical path.

Figure 6.3: Kitchen Remodel, Detail for One Week

Project/Phase/Activity List	Duration	Start	End
The Kitchen	31.00 d	Jan/25/96	Mar/08/96
Demo & Removal	13.00 h	Mar/04/96	Mar/05/96
Dispose of Cabinets	2.00 h	Mar/05/96	Mar/05/96
Remove Existing Cabinets	6.00 h	Mar/04/96	Mar/05/96
Disc and Remove Appliance	4.00 h	Mar/04/96	Mar/04/96
Disc and Remove Sink	1.00 h	Mar/04/96	Mar/04/96
Remove Floor as required	2.00 h	Mar/05/96	Mar/05/96
Patching and Repair	1.63 d	Mar/07/96	Mar/08/96
Replace Flooring	1.00 d	Mar/07/96	Mar/08/96
Repair Wallpaper	3.00 h	Mar/07/96	Mar/07/96
Repair Walls	2.00 h	Mar/07/96	Mar/07/96
New Work	31.00 d	Jan/25/96	Mar/08/96
Owner Select Cabinets	1.00 d	Jan/25/96	Jan/25/96
Order New Cabinets	1.00 d	Jan/26/96	Jan/26/96
Build and Deliver Cabinets	24.00 d	Jan/29/96	Mar/01/96
Install Cabinets	1.50 d	Mar/05/96	Mar/07/96
Install Countertops	4.00 h	Mar/07/96	Mar/07/96
Replace Appliances	2.00 h	Mar/08/96	Mar/08/96
Install Sink	2.00 h	Mar/07/96	Mar/07/96

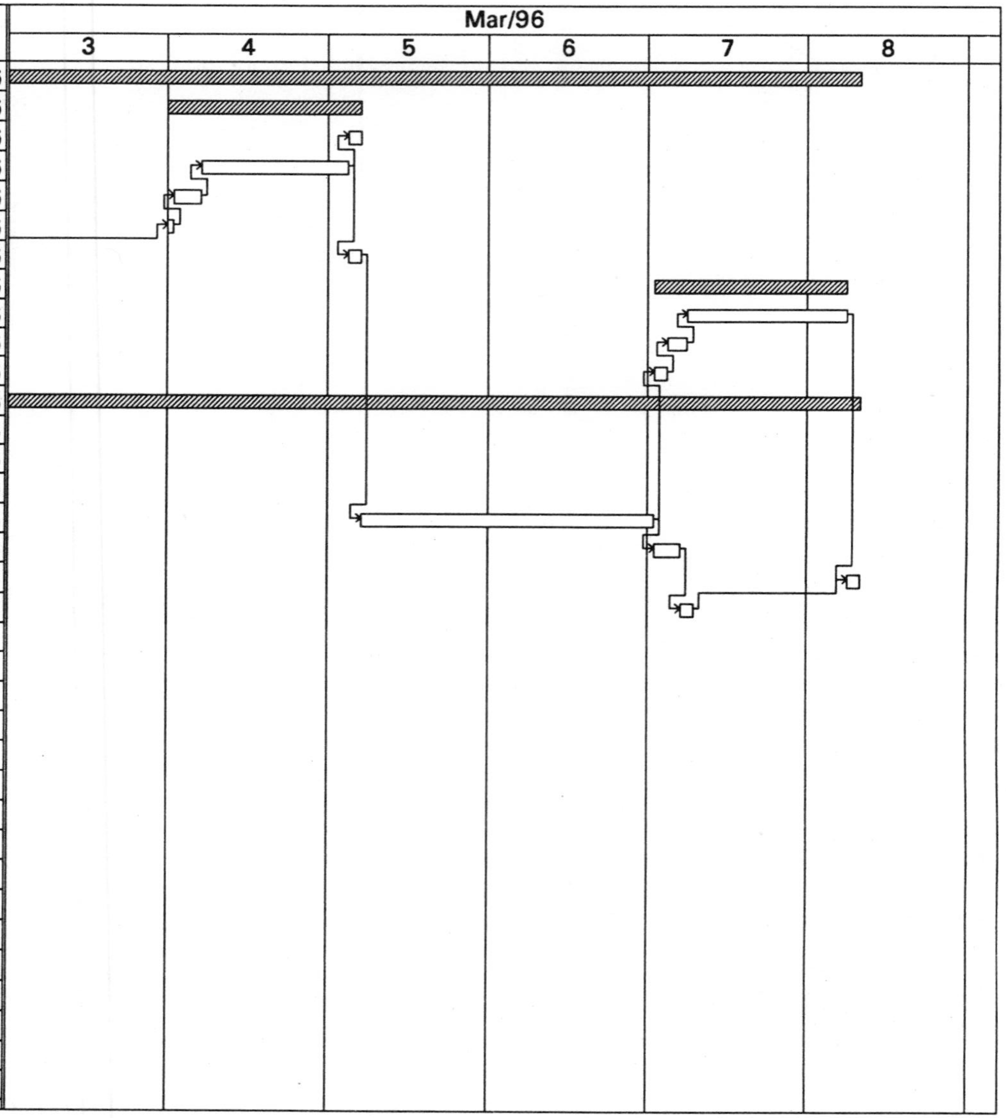

If a subcontractor is used, the remodeler should coordinate with the sub to be sure the method selected by the subcontractor does not create any unnecessary damage.

The plan to complete the addition work is presented in Figure 6.4.

Figure 6.4: Plan for Addition

Activity	Duration, Responsibility
Demolition and Removal	
Remove Hip Section	1 day, Remodeler
Protect Existing Structure	2 hrs., Remodeler
Bring Tarp to Cover Roof	1 hr., Remodeler
Remove Exterior Masonry	1 day, Remodeler
Remove Fascia and Soffits	1 day, Remodeler
Remove Landscaping	2 hrs., Remodeler
Remove Gable Window	6 hrs., Remodeler
Patch and Rework	
Patch at Gable Window	3 hrs., Remodeler
New Door Cut-out	4 hrs., Remodeler
New Work	
Install Skylights in Bath	1 day, Remodeler
Install Slab-on-Grade for Addition	5 days, Concrete Sub
Erect Wall Framing	4 hrs., Remodeler
Erect Roof Framing and Deck	1 day, Remodeler
Place New Shingles and Tie-in Existing	1 day, Roofer
Install New Fascia & Soffits and Tie-in	2 days, Remodeler
Install New Masonry and Tie-in	3 days, Mason
Install Exterior Doors and Windows	3 hrs., Remodeler
Install Interior Trim and Door	4 hrs., Remodeler
Install Drywall	3 days, Remodeler
Paint New Work	2 days, Remodeler
Install New Flooring	4 hrs., Floor Sub

The schedule for the addition is diagrammed in Figure 6.5.

We intend to demolish and remove only the masonry and the landscaping where necessary to install the slab. If desired, we could schedule the slab in greater detail. A more detailed schedule would be similar to the schedule for the slab portion of the house example.

We have elected to complete the slab and new framing before beginning any rework of the existing structure. This means we need to be especially concerned with protecting the existing finish work only for a minimum amount of time. It also keeps the affected area out of service for a short amount of time. Figure 6.5 shows that the "Protect Existing Structure" activity extends for a period of fourteen *workdays*, but twenty-one *calendar* days. We will assume the tarp is removed to place the shingles and will no longer be needed after the shingles are installed. Of course, until the shingles are installed, we must be sure to protect the existing structure.

Figure 6.5: Schedule for Addition

Project/Phase/Activity List	Duration	Start	End
The Addition	20.75 d	Jan/25/96	Feb/23/96
Demolition and Removal	14.13 d	Jan/25/96	Feb/14/96
Bring Tarp to Cover Roof	1.00 h	Jan/25/96	Jan/25/96
Protect Existing Structure	14.00 d	Jan/25/96	Feb/14/96
Remove Landscaping	2.00 h	Jan/25/96	Jan/25/96
Remove Exterior Masonry	1.00 d	Jan/25/96	Jan/26/96
Remove Hip Section	1.00 d	Feb/06/96	Feb/07/96
Remove Fascia and Soffits	1.00 d	Feb/05/96	Feb/06/96
Remove Gable Window	6.00 h	Feb/06/96	Feb/07/96
Patch and Rework	2.75 d	Feb/07/96	Feb/12/96
Patch at Gable Window	3.00 h	Feb/07/96	Feb/07/96
New Door Cut-out	4.00 h	Feb/09/96	Feb/12/96
Replace Hip Section	2.00 d	Feb/07/96	Feb/09/96
New Work	19.50 d	Jan/26/96	Feb/23/96
Inst Slab-on-Grade for Addition	5.00 d	Jan/26/96	Feb/02/96
Erect Wall Framing	4.00 h	Feb/02/96	Feb/02/96
Erect Roof Framing and Deck	1.00 d	Feb/02/96	Feb/05/96
Install Skylights in bath	1.00 d	Feb/09/96	Feb/12/96
Place New Shingles and Tie-in	1.00 d	Feb/13/96	Feb/14/96
Inst New Fascia & Soffits & Tie-in	2.00 d	Feb/09/96	Feb/13/96
Inst Ext. Doors and Windows	3.00 h	Feb/05/96	Feb/06/96
Install New Masonry and Tie-in	3.00 d	Feb/06/96	Feb/09/96
Install Drywall	3.00 d	Feb/14/96	Feb/20/96
Install Interior Trim and Door	4.00 h	Feb/20/96	Feb/21/96
Paint New Work	2.00 d	Feb/21/96	Feb/23/96
Install New Flooring	4.00 h	Feb/23/96	Feb/23/96

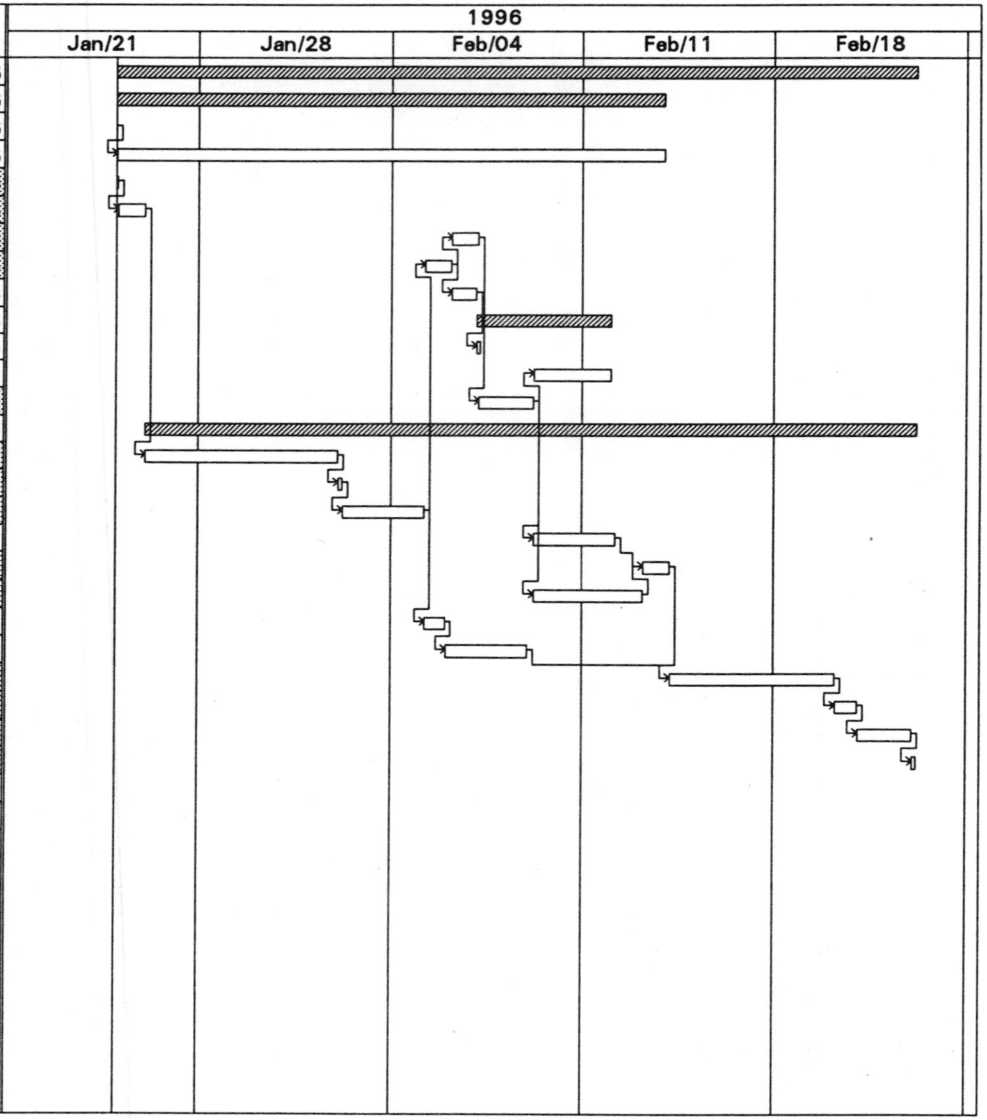

Shaded activities are on the critical path.

Notice that the Protect Existing Structure activity has no logic relationship leaving it. Basically, protection of the structure will be permanent after the shingles have been installed. We could insert a relationship to tie the completion of this activity to the finish of another activity, but we have chosen not to because doing so provides little additional information and clutters the diagram. However, if a delay occurs, the remodeler will have to adjust this activity manually. Be careful if you choose to omit relationships.

Once the new roof framing is in place, the existing hip section of roof, fascia and soffit, and gable window are removed. The hip section is replaced and the skylights and the new fascia and soffit are installed. Also, the patching is complete where the gable window was open to the bathroom area. The exterior doors and windows and the masonry can be started on the exterior of the addition. Once the skylights and fascia are complete, installation of the shingles can begin. After the shingles and masonry work is finished, the drywall in the new section can be started. The door cut-out to the existing bathroom should also be completed. After the drywall, the interior trim, painting, and new flooring are finished.

The start date has been set for January 25. We will adjust the start date once we determine the best overall schedule. Note that the bathroom area will be out of service for more than the one-week (five workday) period specified by the owner. (The bathroom will be closed from the start of Remove Hip Section, February 6, until the skylights are complete on February 12.) To complete the work during the required five workdays, changes to the logic of the schedule and/or duration adjustments or weekend work may be required. If the remodeler believes that this is the best schedule, the owner should be notified before starting this work. Hopefully, the owner will accept the judgment of the remodeler and allow the adjustment.

The Exterior Schedule

The last group of activities deals with the exterior work. Installation of the new garage doors will not create any time problem. The fence will be installed after the new addition is finished. The landscaper will not start until all other work is completed.

The plan for the exterior activities appears in Figure 6.6.

Figure 6.6: Plan for Exterior Remodel

Activity	Duration, Resource
Garage Doors	
Order Garage Doors	1 day, Remodeler
Deliver Garage Doors	9 days, Remodeler
Install Garage Doors	2 days, Remodeler
Modify Opening for Doors	2 days, Remodeler
Exterior Fence	
Set Posts	1 day, Remodeler
Install Fence	2 days, Remodeler
Landscaping	
New Landscaping at Courtyard	3 days, Landscaper
New Landscaping at Addition	2 days, Landscaper
Repair Landscaping	1 day, Landscaper

The schedule diagram for the exterior activities appears in Figure 6.7.

This simple schedule contains only the relationships between the activities shown. These activities have soft restraints to activities in *other* schedules because of the information given above. The times for these activities will change when the overall schedule brings all the activities together into a single diagram.

COMBINING THE SCHEDULES

The next step is to combine our three schedules. Figure 6.8 shows the overall schedule for the entire remodeling and addition work. Using the computer, the schedules are easily combined. The remodeler can perform this task manually; however, it will probably take several rough drafts before a final diagram is completed.

We must look at the overall schedule to determine and resolve the following issues:

- Do we have enough manpower to staff the job to do what is planned?
- Can the subcontractors show up on the days scheduled?
- Do we have to plan for any weather delays or lost productivity because of limited working space?
- Can the owner live with the activities taking place as required?

As shown in the combined schedule, the addition and the exterior work will be completed first. Both areas should be finished by the time the cabinets arrive. The kitchen work will be the last phase to be completed. There is a one-week period starting February 25 when no activities are listed as occurring on the project.

The remodeler might use this time as a buffer to allow for delays in the addition work. However, the owner will have to deal with the disruption of the house from January 25 until the work is completed on March 8. If the remodeler delayed the start of the addition and exterior for three weeks, what impact would this have on the overall project? Figure 6.9 shows the schedule if the addition and exterior work started approximately three weeks later, on Tuesday, February 12.

This new schedule shows that the majority of the work will occur during a three-week period from February 18 to March 12. The remodeler still has a conflict with the owner's wishes concerning the bathroom being out of service. It appears that the remodeler will be working in this area from February 23, with the removal of the hip section, until February 29, when the patching and skylights are complete. The remodeler could use the techniques discussed concerning crashing the schedule or could stay with this schedule. Currently, the removal of the hip is occurring over a weekend time period. This would expose a lot of finish work to possible weather damage. If the remodeler delayed this work until Tuesday, February 26, the problem could be avoided and the schedule for the addition would look like Figure 6.10.

Making this adjustment appears to have helped with the one-week time frame required by the owner, reduced the weather exposure problem, and had no impact on the overall schedule.

The schedule also helps the remodeler to identify critical activities. Critical activities in the kitchen project include the selection, order, delivery, and installation of the cabinets. The repair of the walls and floor are critical at the end of this section. If a problem occurs, the remodeler has a small amount of excess time to adjust. However, the cabinets have to be on time or a delay is sure to occur.

Figure 6.7: Schedule for Exterior Remodel

Phase/Activity List	Duration	Start	End
Garage Door Work	12.00 d	Jan/25/96	Feb/09/96
Order Garage Doors	1.00 d	Jan/25/96	Jan/25/96
Deliver Garage Doors	9.00 d	Jan/26/96	Feb/07/96
Modify Opening for Doors	2.00 d	Jan/25/96	Jan/26/96
Install Garage Doors	2.00 d	Feb/08/96	Feb/09/96
Fence Work	3.00 d	Jan/25/96	Jan/29/96
Set Posts	1.00 d	Jan/25/96	Jan/25/96
Install Fence	2.00 d	Jan/26/96	Jan/29/96
Landscaping	6.00 d	Jan/30/96	Feb/06/96
Landscaping at Courtyard	3.00 d	Jan/30/96	Feb/01/96
Landscaping at Addition	2.00 d	Feb/02/96	Feb/05/96
Repair Landscaing	1.00 d	Feb/06/96	Feb/06/96

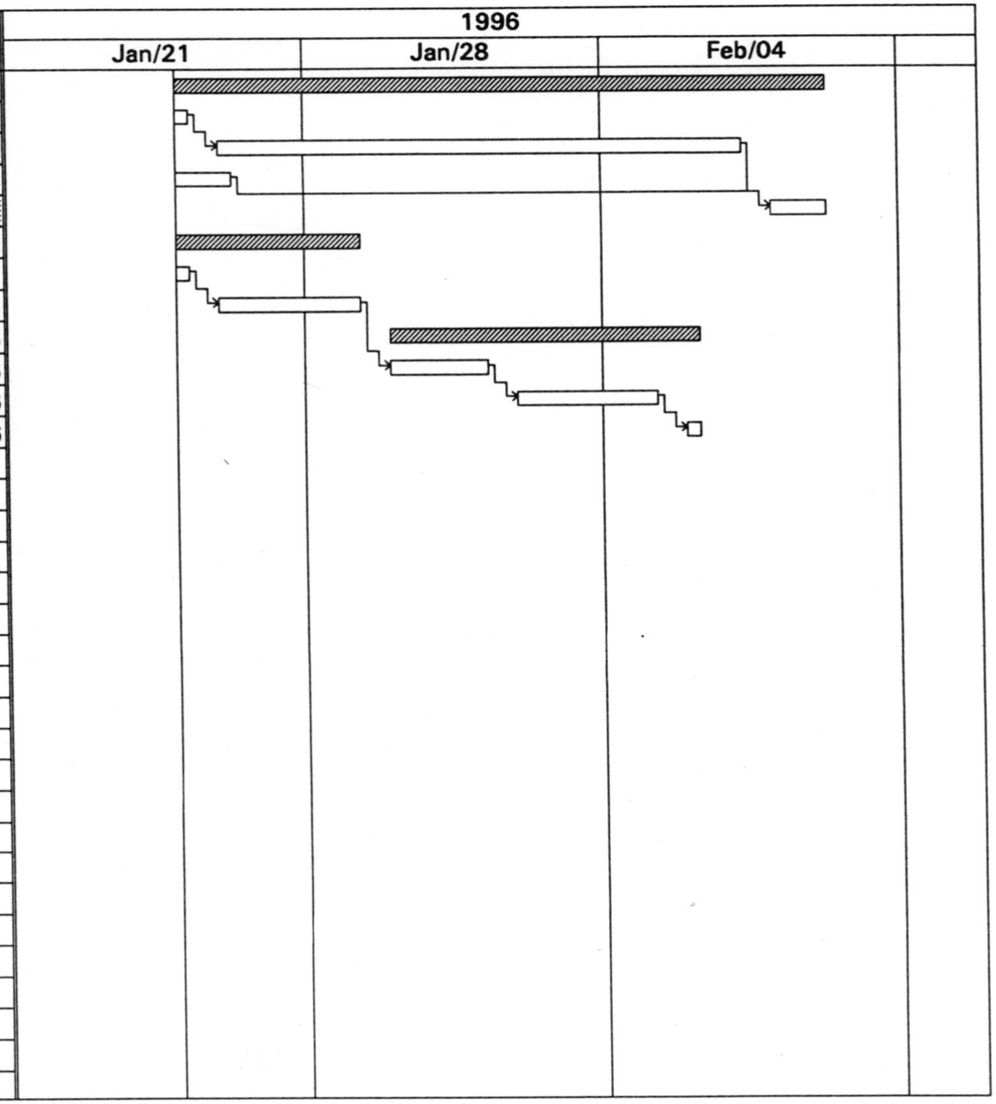

Shaded activities are on the critical path.

Figure 6.8: Master Schedule for Remodel

Project/Activity List	Duration	Start	End
The Kitchen	31.00 d	Jan/25/96	Mar/08/96
Dispose of Cabinets	2.00 h	Mar/05/96	Mar/05/96
Remove Existing Cabinets	6.00 h	Mar/04/96	Mar/05/96
Disc and Remove Appliances	4.00 h	Mar/04/96	Mar/04/96
Disc and Remove Sink	1.00 h	Mar/04/96	Mar/04/96
Remove Floor as required	2.00 h	Mar/05/96	Mar/05/96
Replace Flooring	1.00 d	Mar/07/96	Mar/08/96
Repair Wallpaper	3.00 h	Mar/07/96	Mar/07/96
Repair Walls	2.00 h	Mar/07/96	Mar/07/96
Owner Select Cabinets	1.00 d	Jan/25/96	Jan/25/96
Order New Cabinets	1.00 d	Jan/26/96	Jan/26/96
Build and Deliver Cabinets	24.00 d	Jan/29/96	Mar/01/96
Install Cabinets	1.50 d	Mar/05/96	Mar/07/96
Install Countertops	4.00 h	Mar/07/96	Mar/07/96
Replace Appliances	2.00 h	Mar/08/96	Mar/08/96
Install Sink	2.00 h	Mar/07/96	Mar/07/96
The Addition	20.75 d	Jan/25/96	Feb/23/96
Bring Tarp to Cover Roof	1.00 h	Jan/25/96	Jan/25/96
Protect Existing Structure	14.00 d	Jan/25/96	Feb/14/96
Remove Landscaping	2.00 h	Jan/25/96	Jan/25/96
Remove Exterior Masonry	1.00 d	Jan/25/96	Jan/26/96
Remove Hip Section	1.00 d	Feb/06/96	Feb/07/96
Remove Fascia and Soffits	1.00 d	Feb/05/96	Feb/06/96
Remove Gable Window	6.00 h	Feb/06/96	Feb/07/96
Patch at Gable Window	3.00 h	Feb/07/96	Feb/07/96
New Door Cut-out	4.00 h	Feb/09/96	Feb/12/96
Replace Hip Section	2.00 d	Feb/07/96	Feb/09/96
Inst Slab-on-Grade for Additio	5.00 d	Jan/26/96	Feb/02/96
Erect Wall Framing	4.00 h	Feb/02/96	Feb/02/96
Erect Roof Framing and Deck	1.00 d	Feb/02/96	Feb/05/96
Install Skylights in bath	1.00 d	Feb/09/96	Feb/12/96
Place New Shingles and Tie-in	1.00 d	Feb/13/96	Feb/14/96
Inst New Fascia & Soffits & T	2.00 d	Feb/09/96	Feb/13/96
Inst Ext. Doors and Windows	3.00 h	Feb/05/96	Feb/06/96
Install New Masonry and Tie-i	3.00 d	Feb/06/96	Feb/09/96
Install Drywall	3.00 d	Feb/14/96	Feb/20/96
Install Interior Trim and Door	4.00 h	Feb/20/96	Feb/21/96
Paint New Work	2.00 d	Feb/21/96	Feb/23/96
Install New Flooring	4.00 h	Feb/23/96	Feb/23/96
Garage Door Work	12.00 d	Jan/25/96	Feb/09/96
Fence Work	3.00 d	Jan/25/96	Jan/29/96
Landscaping	6.00 d	Jan/30/96	Feb/06/96

Note: The detailed activities are not shown for the Garage Door Work, Fence Work and Landscaping.

Figure 6.9: Adjusted Schedule for Remodel

Project/Activity List	Duration	Start	End
The Kitchen	31.00 d	Jan/25/96	Mar/08/96
Dispose of Cabinets	2.00 h	Mar/05/96	Mar/05/96
Remove Existing Cabinets	6.00 h	Mar/04/96	Mar/05/96
Disc and Remove Appliances	4.00 h	Mar/04/96	Mar/04/96
Disc and Remove Sink	1.00 h	Mar/04/96	Mar/04/96
Remove Floor as required	2.00 h	Mar/05/96	Mar/05/96
Replace Flooring	1.00 d	Mar/07/96	Mar/08/96
Repair Wallpaper	3.00 h	Mar/07/96	Mar/07/96
Repair Walls	2.00 h	Mar/07/96	Mar/07/96
Owner Select Cabinets	1.00 d	Jan/25/96	Jan/25/96
Order New Cabinets	1.00 d	Jan/26/96	Jan/26/96
Build and Deliver Cabinets	24.00 d	Jan/29/96	Mar/01/96
Install Cabinets	1.50 d	Mar/05/96	Mar/07/96
Install Countertops	4.00 h	Mar/07/96	Mar/07/96
Replace Appliances	2.00 h	Mar/08/96	Mar/08/96
Install Sink	2.00 h	Mar/07/96	Mar/07/96
The Addition	20.75 d	Feb/12/96	Mar/12/96
Bring Tarp to Cover Roof	1.00 h	Feb/12/96	Feb/12/96
Protect Existing Structure	14.00 d	Feb/13/96	Mar/04/96
Remove Landscaping	2.00 h	Feb/12/96	Feb/12/96
Remove Exterior Masonry	1.00 d	Feb/12/96	Feb/13/96
Remove Hip Section	1.00 d	Feb/23/96	Feb/26/96
Remove Fascia and Soffits	1.00 d	Feb/22/96	Feb/23/96
Remove Gable Window	6.00 h	Feb/23/96	Feb/26/96
Patch at Gable Window	3.00 h	Feb/26/96	Feb/26/96
New Door Cut-out	4.00 h	Feb/28/96	Feb/29/96
Replace Hip Section	2.00 d	Feb/26/96	Feb/28/96
Inst Slab-on-Grade for Addition	5.00 d	Feb/13/96	Feb/21/96
Erect Wall Framing	4.00 h	Feb/21/96	Feb/21/96
Erect Roof Framing and Deck	1.00 d	Feb/21/96	Feb/22/96
Install Skylights in bath	1.00 d	Feb/28/96	Feb/29/96
Place New Shingles and Tie-in	1.00 d	Mar/01/96	Mar/04/96
Inst New Fascia,Soffits & Tie-in	2.00 d	Feb/28/96	Mar/01/96
Inst Ext. Doors and Windows	3.00 h	Feb/22/96	Feb/23/96
Install New Masonry and Tie-in	3.00 d	Feb/23/96	Feb/28/96
Install Drywall	3.00 d	Mar/04/96	Mar/07/96
Install Interior Trim and Door	4.00 h	Mar/07/96	Mar/08/96
Paint New Work	2.00 d	Mar/08/96	Mar/12/96
Install New Flooring	4.00 h	Mar/12/96	Mar/12/96
Garage Door Work	17.75 d	Feb/12/96	Mar/07/96
Order Garage Doors	1.00 d	Feb/12/96	Feb/12/96
Deliver Garage Doors	9.00 d	Feb/13/96	Feb/26/96
Modify Opening for Doors	2.00 d	Mar/01/96	Mar/05/96
Install Garage Doors	2.00 d	Mar/05/96	Mar/07/96
Fence Work	3.00 d	Feb/28/96	Mar/04/96
Set Posts	1.00 d	Feb/28/96	Feb/29/96
Install Fence	2.00 d	Feb/29/96	Mar/04/96
Landscaping	6.00 d	Mar/04/96	Mar/12/96
Landscaping at Courtyard	3.00 d	Mar/04/96	Mar/07/96
Landscaping at Addition	2.00 d	Mar/07/96	Mar/11/96
Repair Landscaping	1.00 d	Mar/11/96	Mar/12/96

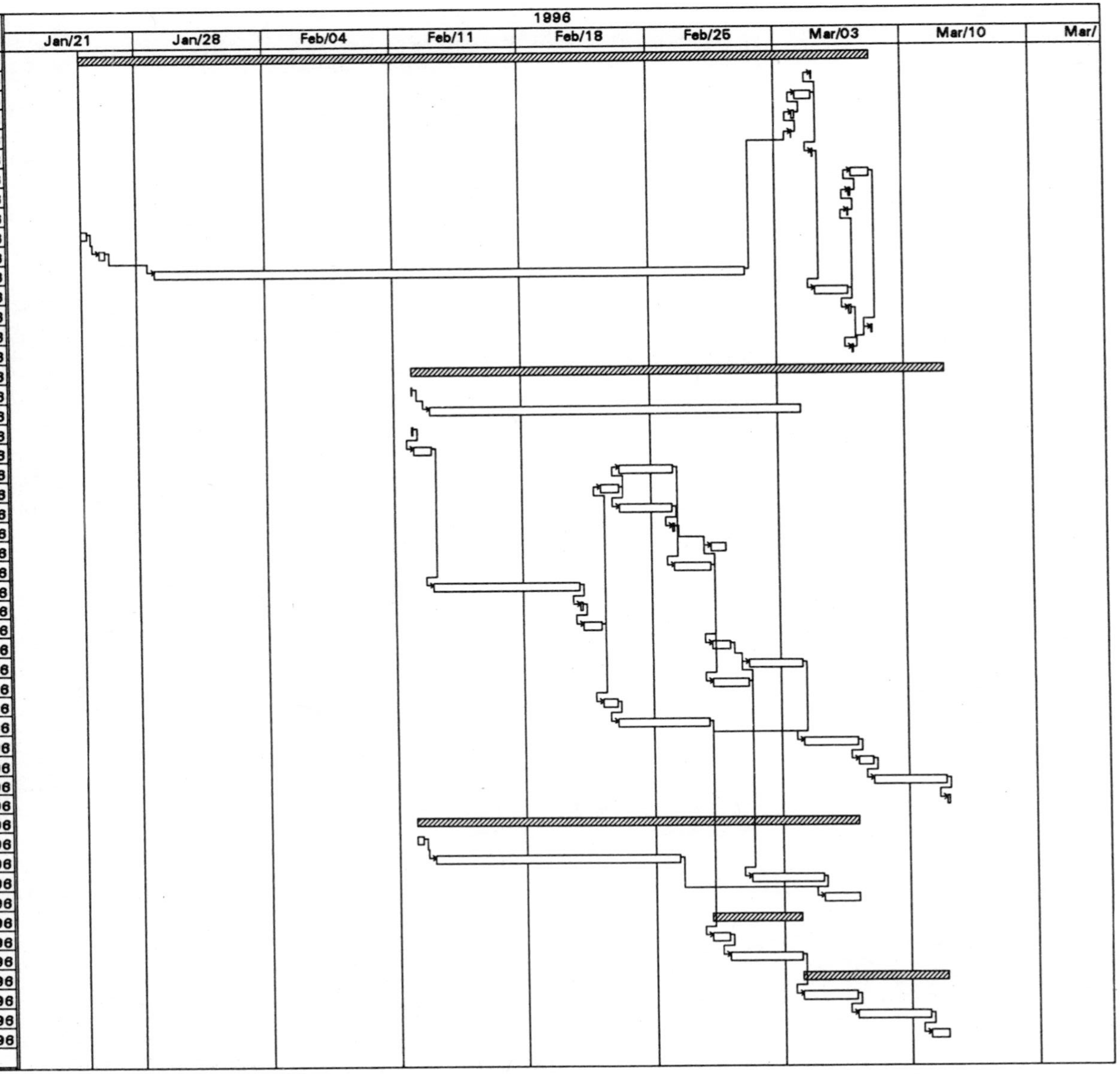

Figure 6.10: Adjusted Schedule for Addition

Project/Activity List	Duration	Start	End
The Addition	20.75 d	Feb/13/96	Mar/13/96
Bring Tarp to Cover Roof	1.00 h	Feb/13/96	Feb/13/96
Protect Existing Structure	14.00 d	Feb/13/96	Mar/05/96
Remove Landscaping	2.00 h	Feb/13/96	Feb/13/96
Remove Exterior Masonry	1.00 d	Feb/13/96	Feb/14/96
Remove Hip Section	1.00 d	Feb/26/96	Feb/27/96
Remove Fascia and Soffits	1.00 d	Feb/23/96	Feb/26/96
Remove Gable Window	6.00 h	Feb/26/96	Feb/27/96
Patch at Gable Window	3.00 h	Feb/27/96	Feb/27/96
New Door Cut-out	4.00 h	Feb/29/96	Mar/01/96
Replace Hip Section	2.00 d	Feb/27/96	Feb/29/96
Inst Slab-on-Grade for Addition	5.00 d	Feb/14/96	Feb/22/96
Erect Wall Framing	4.00 h	Feb/22/96	Feb/22/96
Erect Roof Framing and Deck	1.00 d	Feb/22/96	Feb/23/96
Install Skylights in bath	1.00 d	Feb/29/96	Mar/01/96
Place New Shingles and Tie-in	1.00 d	Mar/04/96	Mar/05/96
Inst New Fascia,Soffits & Tie-in	2.00 d	Feb/29/96	Mar/04/96
Inst Ext. Doors and Windows	3.00 h	Feb/23/96	Feb/26/96
Install New Masonry and Tie-in	3.00 d	Feb/26/96	Feb/29/96
Install Drywall	3.00 d	Mar/05/96	Mar/08/96
Install Interior Trim and Door	4.00 h	Mar/08/96	Mar/11/96
Paint New Work	2.00 d	Mar/11/96	Mar/13/96
Install New Flooring	4.00 h	Mar/13/96	Mar/13/96

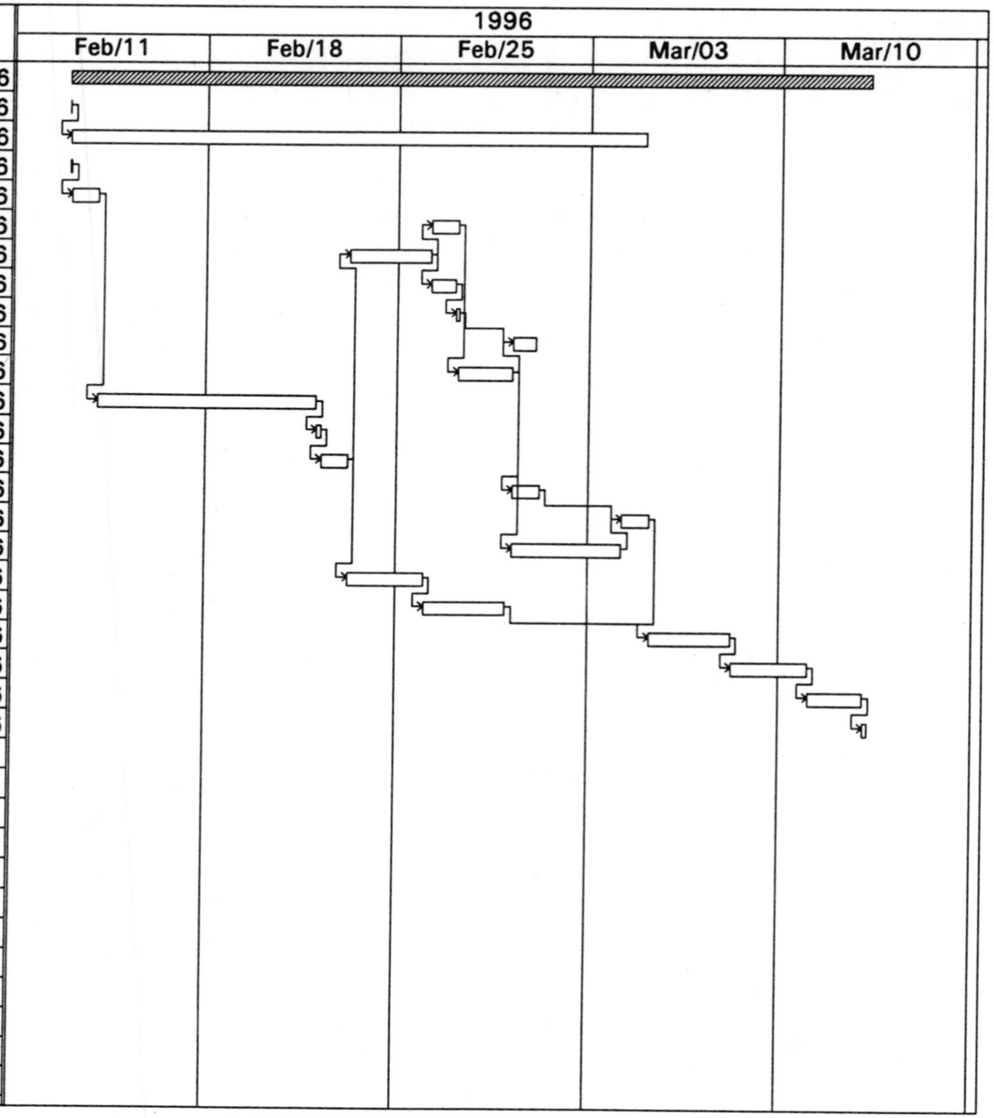

The addition area schedule contains critical activities as well, namely the removal of the landscaping and masonry, the slab installation, the framing activities, the removal of the hip, the repair of the bathroom, the fascia and soffit, the shingles, and the drywall and interior finishes. As this list suggests, most of the activities in this area and in the overall project are critical. Whenever a lot of work has to be completed in a short period of time, most of the activities will be critical.

The exterior activities do not appear to be as critical. These activities can be completed during the time the balance of the project is being completed. However, since we moved these activities to a later stage in the project, a delay for weather or some other reason could cause a project delay. We can use the same techniques described in the house example to adjust scheduled activities in case of weather delays.

This combined schedule contains information that the remodeler, subcontractors, and homeowner can all use. If all goes according to plan, the project will be a success for everyone. Tracking the progress of the work will be important. Remodeling projects with short-duration activities and short project durations will have to be monitored frequently, such as every day or every two days.

The update process is the same for any schedule developed. Inspections, determining remaining durations, and recalculating the schedule are required. This can be time-consuming if the update is performed manually every day. A computerized scheduling program is a must if the remodeler is going to properly track the project.

Remember, communication, cooperation, and teamwork will be important as well. The time spent in creating the schedule can save time on the project if everyone is ready to work.

SUMMARY POINTS AND TIPS

- Use the estimate and drawings to determine the plan to complete the work.
- Remodelers must work around existing construction and often must take special action to protect all or a portion of the existing structure. The schedule must reflect these constraints and identify all potential problems.
- Breaking the work down into sections or by locations is sometimes helpful. The specific schedules can then be combined to review the relationships between activities in different areas.
- All of the techniques discussed in the other chapters can be incorporated into remodeling schedules. For example, cash-flow tracking and resource leveling can be tied to any schedule.

Chapter 7

Legal Implications of Scheduling

Builders and remodelers can spend a great deal of time planning and scheduling and still end up with problems. Murphy's law is alive and well in the construction business. Materials are delivered late, subcontractors go out of business, owners request changes in the project, bad weather moves in and won't go away—virtually every project can be beset with complications.

Often, such complications delay the progress of construction. The original schedule can quickly become useless for determining the start times for the remaining activities. The builder or remodeler then uses scheduling, management, and communication skills to make corrections in the workflow.

The needed corrections can be minor or major depending on the problem. Minor changes usually are expected by all parties, and the initial schedule usually makes allowances for some variances. However, major schedule changes can have a negative impact on everyone involved with the project.

The more off-track the schedule becomes, the more builders and remodelers are at risk of action by subcontractors, material suppliers, and the customer. A seriously delayed project generally brings in less income. Subcontractors who have other projects to complete may not be able to accommodate schedule changes. A new home buyer may have sold his or her previous home and need to be in the new home by a certain date. A remodeling customer may become unnerved at delays in a project, particularly if the customer is living in the home while the remodeling work takes place.

When people lose money on projects, they sometimes take their problems to court. The builder, subcontractors, and customers will all be affected by going to court. No matter what led to the problem, no one will leave the project with a good feeling.

Good scheduling, in conjunction with other sound business practices, can help prevent misunderstandings or delays that can land you in court. However, in some situations you may need to use caution in communicating schedule information to people outside your company.

A logic diagram is just a plan showing the order and the time in which the builder plans to do the various activities of a job. In the event of a dispute about time, such as finishing a job later or not being ready for a subcontractor as originally planned, the logic diagram could be used against the builder in a court proceeding under the theory that the logic diagram represents a commitment on the part of the builder to do various segments of the work by certain times. The chance of this happening is rare, but such actions have been brought. Therefore, the builder should take steps for his or her protection in the case of a dispute. The first thing is not to give out copies of detailed scheduling logic diagrams indiscriminately. If it is necessary to hand out a logic diagram, add a note that the diagram is simply a planning guide for the builder and should not be interpreted as a commitment to do anything other than what is in the contract.

REFERENCES IN CONTRACTS

This book has stressed the importance of teamwork, communication, and cooperation among the parties involved in residential construction. A project completed on schedule is the result of everyone working together.

Sooner or later, however, a nightmare project arrives on everyone's doorstep. This is the project that never gets finished, for which nothing goes right, during which weather goes from bad to worse, or on which the workers do everything wrong. Several factors can cause a project to become a nightmare project: a bad schedule (or no schedule at all); poor communication between the builder or remodeler and the subs; unrealistic expectations on the part of owners; changes in project scope; and unforeseen catastrophes.

What can you do to reduce the likelihood of a nightmare project? Communicate exactly what is expected. Use contracts that express both the time and scope of work to be completed as clearly as possible. Contracts are only as good as the people signing them, of course, so be sure the parties can deliver on their obligations. Where a contract has a fixed start and end date, it is up to the builder to manage the activities during the life of the project. Builders should safeguard their flexibility to move people around or change the order of activities so long as this does not delay or interfere with subcontractors' ability to get their work done.

If you share schedule diagrams with owners, you also may wish to include on the schedule document or in the owner's contract a paragraph clarifying the role and importance of the schedule. This paragraph might be similar to the following:

> *The attached schedule represents the builder's original work plan for completing the project. The completion date is (actual date). It is understood by both parties that the work plan may change and that the completion date may change if the schedule is affected by events beyond the builder's control.*

(Note: It is imperative that your legal counsel review any sample language you may consider incorporating into an actual contract.)

Linking the schedule to the contract causes the builder or remodeler, subcontractors, and owners to commit legally to delivering the product on a time schedule. Can the schedule be used against the builder or remodeler who has fallen behind? It sure can. However, if the builder is monitoring the schedule properly, most serious schedule problems should be known in advance. And if the contract language provides the builder or remodeler with the right to reasonably adjust the schedule as needed, the contract language—along with the schedule—can be used to support the builder's or remodeler's position in court should a dispute later arise. For example, just as the responsibility codes assigned to each activity help when communicating during the project, they also provide a record of responsibility should you have to go to court.

Ideally, advance communication with workers and subcontractors about potential problems should prevent major delays. The builder should also remember the schedule is developed to help construct the project. The final completion date can be adjusted to accommodate anticipated weather delays and contingencies by adding time to activities within the schedule. The schedule should be as accurate as possible to help motivate and control the work. However, committing to a schedule without some contingent time available is not a good business practice.

The builder or remodeler should develop a scheduling system that includes realistic durations, regular updating periods, and ongoing communication. If this is done, the schedule should provide the builder or remodeler with an early warning system that can help prevent legal battles.

Possible Liabilities

A logic diagram is just a plan showing the order and the time in which the builder plans to do various activities. In the event of a dispute about time—such as a completion date for a presold house—the schedule *might* be used against the builder or remodeler in a court proceeding under the theory that the schedule represents a commitment to complete various segments of the work by certain times. This is a rare occurrence in residential building and might be avoided if the contract addresses the issue; however, precedents exist from commercial building, and certain builders (for example, custom builders) may be more vulnerable than others to litigation and associated costs arising out of schedule changes or conflicts.

When remodeling projects have delays and problems, claims for damages invariably include costs associated with delays and/or late completion. Historically, remodelers have enjoyed wide latitude in constructing a defense against the delay claims. For example—"it was the owner's fault since he failed to do something," "he changed something," or "a hidden existing defect caused a problem." If the remodeler provides a written schedule, the owner may give a copy to an attorney if a problem arises. A good claims attorney may be able to use the schedule against the remodeler. Conversely, however, a smart remodeler can use a good schedule to document actual delays caused by hidden defects or owner changes and help defend the remodeler's claims for additional costs.

Potential liability to subcontractors may arise from:

- Logic changes. If the schedule shows a certain logic and the builder or remodeler changes sequencing, the subcontractor may claim its work is more difficult and costly.
- Timeframes. If the schedule shows a certain timeframe and the builder or remodeler has to adjust start times, the subcontractor may claim extra costs.
- Duration changes. If the schedule indicates a certain amount of time for an activity and the builder or remodeler requires the work to be completed more quickly, a subcontractor may claim extra costs due to acceleration of the work schedule.

These issues can sometimes be addressed through contract language. Even with such language, however, builders and remodelers may be liable if the schedule adjustments they make are deemed to be extreme.

Potential liability to owners may arise on certain projects from:

- Project delays. The owner may claim additional costs such as hotel expenses, interest payments, and other costs if the project is not completed according to the contract date.
- Denial of time extension: The owner may refuse to grant time extensions for unexcused delays in activities off the critical path.

Many of these situations may never occur. How can you guard against these possibilities? Using a sound planning, scheduling, and monitoring system will help you avoid trouble. The disciplined scheduling practices, high-quality information, and improved communication that can result from consistent use of CPM scheduling make it less likely that disputes will escalate into litigation.

Problems and changes will still occur. Their impact on a project can be minimized if everyone works together to solve the problem instead of using it as an excuse to delay the project or avoid doing the work.

Finding and maintaining a good team of subcontractors is not an easy task. Builders and remodelers must invest a considerable amount of trust and cooperation in relationships with key subcontractors to obtain a good product. If this trust and commitment is a two-way street, the team can work properly. The builder or remodeler can then consult with the subcontractors when the schedule is prepared and expect everyone to stick as closely as possible to the schedule.

SUMMARY POINTS AND TIPS

- To avoid potential legal problems associated with the schedule, work at communicating with the subcontractors, owners, and other people on the team. Improving communication does not happen automatically; builders and remodelers must make an effort to share needed information, set goals, listen to feedback from the other parties, and reconcile varying priorities.
- Parties to any contract referencing a schedule should discuss the schedule *before* the contract is signed. Before asking them to commit, encourage subcontractors to communicate their views as to how the project will best be completed. When changes or delays occur during the project, be forthright and give other responsible parties enough time to make adjustments. Also, if an event occurs that will cause a project delay, the builder or remodeler should always give written notice to the owner and request an extension of time corresponding to the delay.
- Explain your scheduling system to the subcontractors involved on the project. The subcontractor—who may be working with various builders or remodelers, each using a different system—may not like having to plan the work in detail or commit to the future date. However, you will both gain a lot of advantages if a sound scheduling system is used properly. Be patient and be willing to compromise in noncritical areas to win over the subcontractor to your system. Once the subcontractor sees the benefits of working with you this way, you will have a long-term ally.
- Customers look for builders and remodelers who can deliver a product in a professional manner. You look professional when you furnish a time-scaled bar chart similar to the examples shown in this book. Timely, accurate communication with customers regarding the schedule gives a customer confidence in your ability to manage situations as they come up. Obviously, customers' first preference is to have the job completed on the date originally planned. If that is not possible, however, their next greatest concern is to be reassured that you are handling any problems efficiently and considering their comfort and investment. Telling the customer about every little problem is overkill, of course, but certainly customers should be alerted to delays that you know will affect the completion date.
- Additional contract language and information on legal subjects can be obtained from the book *Contracts and Liability for Builders and Remodelers*, third edition, published by the Home Builder Press and the Legal Department of the National Association of Home Builders.

Appendix A: Sample House Plan

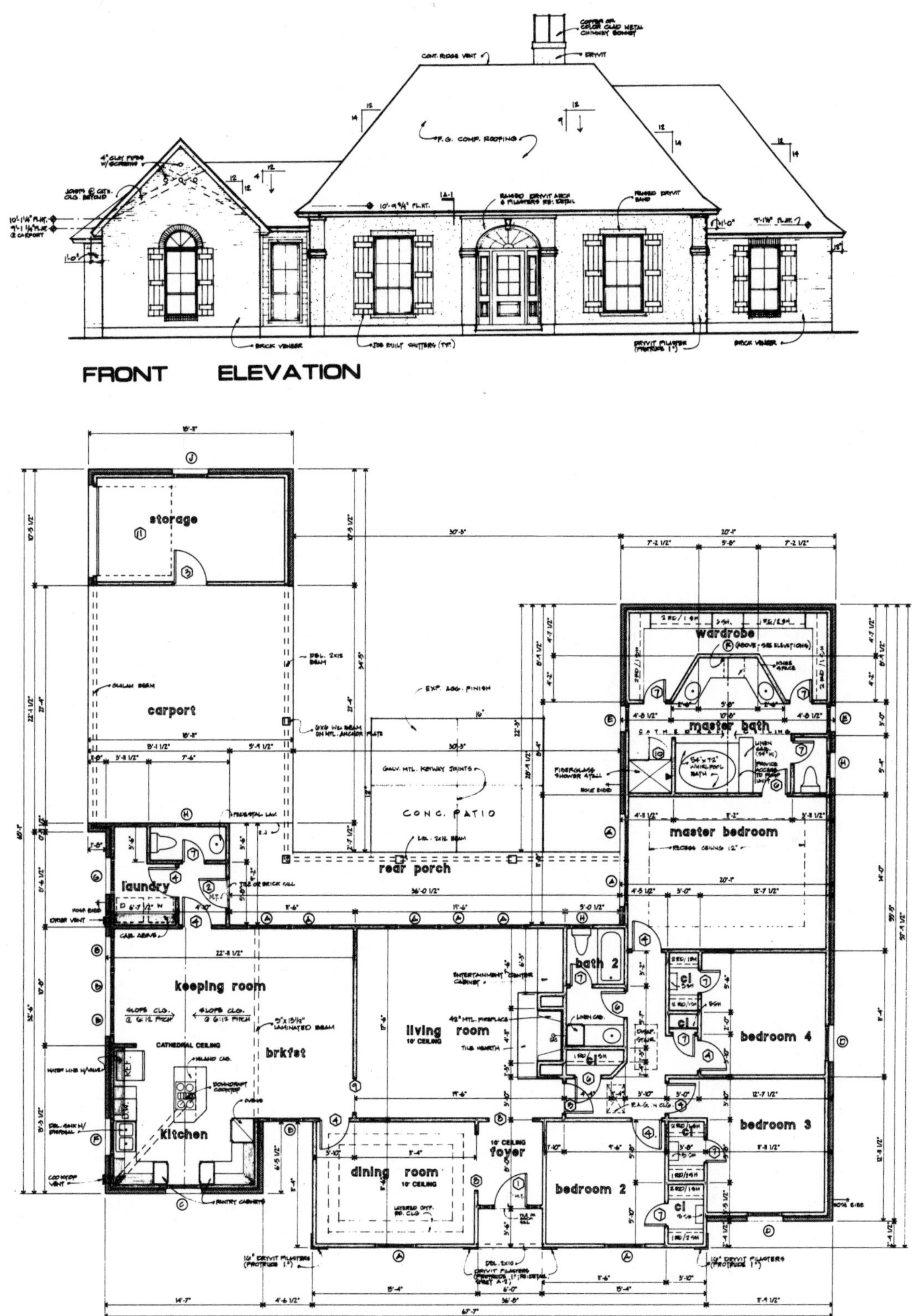

Plans created by David Blackwell, Baton Rouge Plan Service. Used by permission.

Appendix B: Estimate for the Sample House

Thomas A. Love	Estimating Ext Standard Report House				
DESCRIPTION	LABOR AMOUNT	MATRL AMOUNT	SUB AMOUNT	EQUIP AMOUNT	TOTAL AMOUNT
1.000 GEN CONDITIONS					
1.002 Land Costs			61,500		61,500
1.003 Closing Costs			22,250		22,250
1.005 Supervision	3,750				3,750
1.016 Material & Equipment	200	150	250		600
1.017 Misc. Overhead	250	100	1,000		1,350
1.200 Proj. Utilities		200			200
1.500 Permits & Licenses			500		500
1.900 Project Close-out			500		500
GEN CONDITIONS	***4,200***	***450***	***86,000***		***90,650***
2.000 SITE WORK					
2.110 Clear & Grub	1,250			1,450	2,700
2.235 Paving	173	5,047		26	5,246
2.262 Gravel Fill	224	412		20	656
2.300 Excavation - BLDG	310				310
2.528 Sidewalks	360	494			854
2.800 Soil Treatment			1,000		1,000
2.900 Site Improvements			2,500		2,500
SITE WORK	***2,316***	***5,953***	***3,500***	***1,497***	***13,266***
3.000 CONCRETE					
3.180 Slabs on Grade	386	3,525		81	3,992
3.190 Slab Finishing	884			188	1,072
3.210 Reinforcing Steel	140	367			508
3.220 Welded Wire Mesh	154	281			436
3.225 Vapor Barrier	67	67			134
3.500 Formwork	735	347		30	1,112
3.510 Slab Joints	31	37			68
CONCRETE	***2,397***	***4,626***		***298***	***7,322***
4.000 MASONRY					
4.100 Mortar		221			221
4.155 Masonry Accessories		612	790		1,402
4.210 Face Brick		3,606	2,664		6,270
MASONRY		***4,439***	***3,454***		***7,893***

Estimates created using Timberline Precision Estimating Software. The mathematical operation in Precision Estimating rounds the amounts to the nearest dollar. The rounded totals that appear for each major portion of the summary estimate may therefore be slightly higher or lower than the actual sum of the rounded numbers shown in the preceding columns.

Thomas A. Love	Estimating Ext Standard Report House				
DESCRIPTION	LABOR AMOUNT	MATRL AMOUNT	SUB AMOUNT	EQUIP AMOUNT	TOTAL AMOUNT
5.000 METALS					
5.500 Misc. Metals	179	443			621
METALS	***179***	***443***			***621***
6.000 CARPENTRY & MILLWORK					
6.115 Framing Ceiling Systems		1,272	858		2,130
6.120 Framing Roof Systems		1,810	2,368		4,178
6.128 Framing Walls		1,887	2,547		4,434
6.154 Sheathing		1,876	1,121		2,997
6.220 Exterior Wood Trim		92	1,061		1,153
6.225 Exterior Soffits		326	393		719
6.270 Exterior Sidings		317	51		368
6.272 Dryvit Finishes			2,371		2,371
6.290 Exterior Accessories	25	360			385
6.300 Interior Trim		440	848		1,289
6.410 Casework			7,580		7,580
6.760 Exterior Shutters	152	164			316
6.900 Misc. Framing Items		266			266
CARPENTRY & MILLWORK	***177***	***8,811***	***19,198***		***28,185***
7.000 WATERPRF & INSUL					
7.211 Blown-in Insulation		902	575	338	1,814
7.226 Batt Insulation		558	279		837
7.310 Shingles		1,526	1,653		3,179
7.620 Sheet Metal Flashing		611	347		958
7.650 Plastic Thru-Wall Flhin		35			35
WATERPRF & INSUL		***3,631***	***2,854***	***338***	***6,822***
8.000 DOORS & WINDOWS					
8.262 Wood Doors		1,983	505		2,487
8.360 Overhead Doors		275	135		410
8.520 Aluminum Windows		3,186	369		3,555
8.710 Hardware		460	207		667
DOORS & WINDOWS		***5,904***	***1,216***		***7,120***

Thomas A. Love

Estimating Ext Standard Report
House

DESCRIPTION	LABOR AMOUNT	MATRL AMOUNT	SUB AMOUNT	EQUIP AMOUNT	TOTAL AMOUNT
9.000 FINISHES					
9.260 Gypsum Board Systems		2,144	2,466		4,610
9.310 Ceramic Tile			6,784		6,784
9.660 Resilient Tile Flooring			204		204
9.685 Carpet			2,001		2,001
9.910 Exterior Painting			1,550		1,550
9.920 Interior Painting			1,970		1,970
9.970 Wallpaper			2,140		2,140
FINISHES		***2,144***	***17,115***		***19,259***
11.000 EQUIPMENT					
11.100 Appliances	200	2,695			2,895
11.200 Special Equipment	63	360	3,576		3,999
EQUIPMENT	***263***	***3,055***	***3,576***		***6,894***
15.000 MECHANICAL SYSTEMS					
15.100 Plumbing			6,527		6,527
15.500 HVAC			10,041		10,041
MECHANICAL SYSTEMS			***16,568***		***16,568***
16.000 ELECTRICAL					
16.100 Electrical			8,368		8,368
ELECTRICAL			***8,368***		***8,368***

ESTIMATE TOTALS

	9,532	Labor	
	39,454	Material	
	161,848	Subcontractor	
	2,133	Equipment	
212,967			
	745	Builder Risk Insurance	.35000%
	21,371	Overhead	10.00000%
	19,417	Profit	8.25970%
254,500	TOTAL ESTIMATE 76.04/sf		

Appendix C: Sample Kitchen Remodel

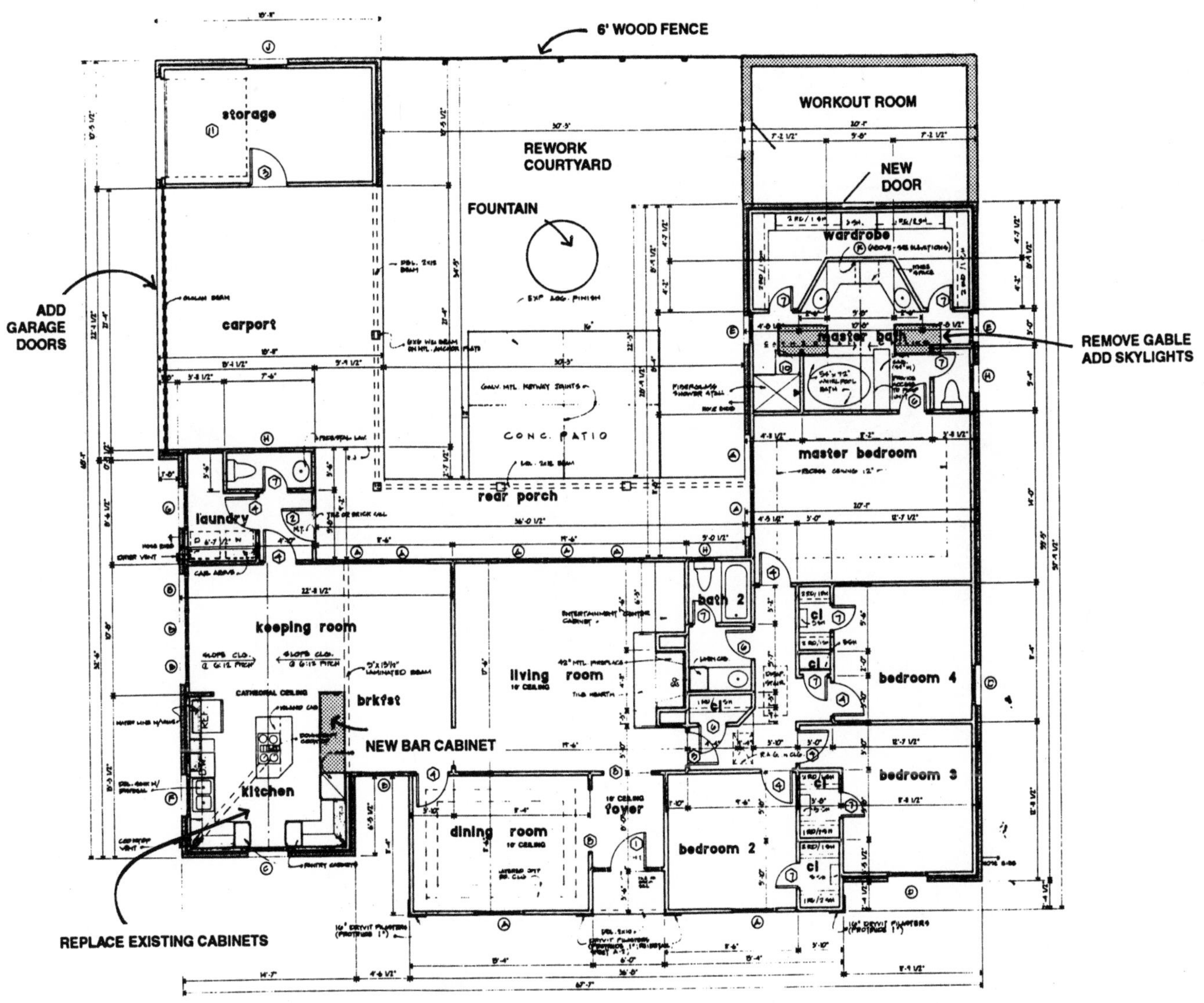

Plans adapted from original house plans by David Blackwell, Baton Rouge Plan Service. Used by permission.

Appendix D: Kitchen Remodel Estimate

Thomas A. Love	Estimating Ext Standard Report Remodel Project				
DESCRIPTION	LABOR AMOUNT	MATRL AMOUNT	SUB AMOUNT	EQUIP AMOUNT	TOTAL AMOUNT
1.000 GEN CONDITIONS					
1.005 Supervision	4,451				4,451
1.500 Permits & Licenses			500		500
1.900 Project Close-out			500		500
GEN CONDITIONS	***4,451***		***1,000***		***5,451***
2.000 SITE WORK					
2.050 Demolition	2,697			38	2,734
2.075 Patching and Rework	1,058	340			1,398
2.085 Protect Existing Work	846	200			1,046
2.830 Fences & Gates	1,269	342			1,611
2.950 Landscape Sub			2,590		2,590
SITE WORK	***5,870***	***882***	***2,590***	***38***	***9,379***
3.000 CONCRETE					
3.180 Slabs on Grade			1,688		1,688
CONCRETE			***1,688***		***1,688***
4.000 MASONRY					
4.010 Masonry Subcontractor			2,450		2,450
MASONRY			***2,450***		***2,450***
6.000 CARPENTRY & MILLWORK					
6.120 Framing Roof Systems	1,964	1,260			3,224
6.128 Framing Walls	327	320			647
6.225 Exterior Soffits	1,310	660			1,970
6.300 Interior Trim	118	126			244
6.410 Casework			12,589		12,589
CARPENTRY & MILLWORK	***3,720***	***2,366***	***12,589***		***18,675***
7.000 WATERPRF & INSUL					
7.310 Shingles			1,898		1,898
7.810 Skylights	846	358			1,204
WATERPRF & INSUL	***846***	***358***	***1,898***		***3,102***

Estimate created using Timberline Precision Estimating Software. The mathematical operation in Precision Estimating rounds the amounts to the nearest dollar. The rounded totals that appear for each major portion of the summary estimate may therefore be slightly higher or lower than the actual sum of the rounded numbers shown in the preceding columns.

Thomas A. Love					
Estimating Ext Standard Report **Remodel Project**					
DESCRIPTION	LABOR AMOUNT	MATRL AMOUNT	SUB AMOUNT	EQUIP AMOUNT	TOTAL AMOUNT
8.000 DOORS & WINDOWS					
8.100 Metal Insulated Doors		120			120
8.262 Wood Doors	26	70			96
8.375 Garage Doors	1,692	980			2,672
8.520 Aluminum Windows	79	446			525
DOORS & WINDOWS	***1,798***	***1,616***			***3,413***
9.000 FINISHES					
9.260 Gypsum Board Systems			1,575		1,575
9.685 Carpet			1,589		1,589
9.925 Painting Subcontract			1,295		1,295
FINISHES			***4,459***		***4,459***

ESTIMATE TOTALS

	16,684	**Labor**	
	5,222	**Material**	
	26,674	**Subcontractor**	
	38	**Equipment**	
48,618			
	225	**Builder Risk Insurance**	**.46200%**
	7,331	**Overhead**	**15.01000%**
	8,426	**Profit**	**15.00000%**
64,600	**TOTAL ESTIMATE**		

Increase Your Business Knowledge with These Bestsellers

Basic Construction Management: The Superintendent's Job, 3rd Edition

The perfect mini-course in field management for superintendents and small-volume builders who direct all phases of construction. ISBN 0-86718-406-X. $27.50

Contracts and Liability for Builders and Remodelers

Contains valuable contract language that can be inserted directly into your own model contracts. ISBN 0-86718-376-4. $25

Customer Service for Home Builders

Provides proven techniques for operating an effective, yet inexpensive, customer service program regardless of your company's size. ISBN 0-86718-355-1. $15

Estimating for Home Builders

Contains all you need to know to produce complete accurate cost estimates, including information on computerized estimating. ISBN 0-86718-372-1. $28

How to Hire and Supervise Subcontractors

Learn how to work more efficiently with subcontractors. ISBN 0-86718-366-7. $15

Production Checklist for Builders and Superintendents

Use this comprehensive checklist to complete projects on time and within budget. ISBN 0-86718-351-9. $20 book only, $35 with diskette, $25 diskette only.

Software Directory

An annually updated directory of software packages programmed just for home builders. Organized into 11 different categories listing hundreds of software packages. Quick-reference grid lets you compare programs at a glance. $15

Your Business Plan: How to Create It, How to Use It

Let this guide take you through the creation of a strategic business plan. Includes a model business plan based on an actual small-volume building company. ISBN 0-86718-390-X. $22

To order any of these books, or request an up-to-date catalog of Home Builder Press titles, write or call:

Home Builder Bookstore
1201 15th St ,NW
Washington, DC 20005-2800
800-223-2665

NAHB MEMBERS RECEIVE A 20% DISCOUNT ON ALL BOOKS

Prices are subject to change.